Biochemical Adaptation
in
Parasites

Biochemical Adaptation
in
Parasites

CHRISTOPHER BRYANT
Professor of Zoology and Dean of Science
at the Australian National University

and

CAROLYN A. BEHM
Lecturer in Biochemistry
at the Australian National University

(with a chapter by Michael J. Howell)

London   New York
CHAPMAN AND HALL

First published in 1989 by
Chapman and Hall Ltd
11 New Fetter Lane, London EC4P 4EE

Published in the USA by
Chapman and Hall
29 West 35th Street, New York NY 10001

© 1989 C. Bryant and C.A. Behm

Printed in Great Britain by
TJ Press Ltd, Padstow, Cornwall

ISBN 0–412–32530–6

British Library Cataloguing in Publication Data

Bryant, Christopher.
Biochemical adaptation in parasites,
1. Parasites
I. Title II. Behm, Carolyn
574.5'249

ISBN 0–412–32530–6

Library of Congress Cataloging in Publication Data

Bryant, Christopher.
Biochemical adaptation in parasites/by Christopher
Bryant and
Carolyn A. Behm (with a chapter by Michael J. Howell).
p. cm.
Bibliography: p.
Includes index.
ISBN 0–412–32530–6
1. Parasites--Physiology. 2. Adaptation (Physiology)
3. Adaptation (Biology) I. Behm. Carolyn A. II. Howell.
Michael
J. III. Title.
[DNLM: 1. Adaptation, Biological. 2. Adaptation,
Physiological.
3. Parasites--physiology.   QX 4 B915b]
QL757.B79   1989
591.52'49--dc20
DNLM/DLC
for Library of Congress   89–15847
                                              CIP

Contents

Preface

In this book we have set out to do something that has not been attempted for parasites since 1970. In that year, in a review entitled *Biochemical Adaptation and the Loss of Genetic Capacity in Helminth Parasites,* Donald Fairbairn surveyed the biochemical literature on parasites. His object was twofold. First, he wished to see whether patterns of adaptation to the parasitic habit existed. Second, he wished to determine whether the remarkable deviations from conventional metabolism so frequently found in parasitic worms were biochemical examples of the simplifications of form considered to be the normal consequences of the assumption of the parasitic habit. In fact, Fairbairn concluded that he could not find a single unambiguous example of biochemical adaptation to parasitism and ascribed the fascinating differences between host and parasite to processes of epigenesis.

In 1970, however, the study of parasite biochemistry was still in its youth. Certainly there were great practitioners, and all workers in the field owe a debt to one of them, Theodor Von Brand and his mammoth compendium *The Biochemistry of Parasites* . The character of research publications at this time was descriptive. Workers were concerned with charting the coastline of a *terra incognita* and with determining whether there were such large-scale features as mountains, and did not worry much about the underlying structure of the new continent. And even if they had been concerned, parasite material is difficult to work with in the absence of many of the biochemical tools we take for granted today.

Nearly twenty years later, how is the story different? A great deal more work has been done on functional aspects of the biochemistry of parasites. Biochemical regulation has become a recurring theme. The unravelling of immune mechanisms has provided new insights into the host-parasite relationship. And the new techniques of molecular biology are beginning to illuminate evolutionary processes.

It was in this context that we conceived the plan for a book on biochemical adaptation as early as 1979 and, eventually, we were invited to write it. We did not want to write a text-book of biochemistry. Barrett's *Biochemistry of Helminth Parasites,* an excellent synthesis that is surely due for second edition, has made it unnecessary for us to do that, thank goodness! Rather, we have attempted to place what is known about the biochemistry of parasites in a biological context. We are concerned with evolution, with ecology, with adaptation and with variation.

Our first idea was to confine ourselves to helminth parasites. Indeed, when we discussed this with our publishers, we had a plan for a book about the size of this one devoted solely to worms. The publishers made the suggestions that, as there was no recent biochemical text on Protozoa, we should include them and, later, that perhaps we should add a section on immunology.   Protozoa are indeed included, and we have been fortunate enough to persuade Dr Michael Howell to write the chapter on parasite immunology.   We have not, however, attempted to cover in great detail molecular aspects of parasitism.   We consider that it is far too early in the development of the subject to permit a useful synthetic approach, although we are aware that the rate that new discoveries are being made suggests that, if *Biochemical Adaptation in Parasites* proceeds to a second edition, a new section will be needed.

We do, of course, owe our publishers a great debt for their patience.   During the preparation of this book all sorts of vicissitudes befell the writers and the Australian education system in general.   All three of us changed our jobs.   Australian tertiary education underwent major restructuring and the Australian National University embarked on an enforced marriage with a neighbouring institution.   For these reasons and others, *Biochemical Adaptation in Parasites* was a long time a-borning.   During its gestation, it has passed through no fewer than three generations of microcomputers and word-processing programs – Wordstar, Wordplex and, finally, MicrosoftWord!

We are deeply indebted to Ms Wendy Lees who, in between her other duties, produced the typescript without complaint and always sooner than we expected. Our colleagues over the years also deserve our gratitude for providing us with a stimulating forum in which to discuss our ideas.   In particular we would like to thank Eva Bennet, Karen Ovington, Nick Smith, Rod Chevis and Stuart Kohlhagen, much of whose work has found its way into this book.

CB and CAB
Australian National University
Canberra, Australia

Enzymes Discussed in the Text

Enzyme	EC No.
acetaldehyde dehydrogenase	1.2.1.101
acetate CoA transferase	2.8.3.8
acetylcholinesterase	3.1.1.7
acetyl CoA synthetase (ADP)	6.2.1.13
acid phosphatase	3.1.3.2
acyl CoA dehydrogenase	1.3.99.3
acyl CoA transferase	2.8.3.n
adenine phosphoribosyltransferase	2.4.2.7
adenine deaminase	3.5.4.2
adenine nucleosidase	3.2.2.1
adenosine deaminase	3.5.4.4
adenosine kinase	2.7.1.20
adenosine nucleosidase	3.2.2.7
adenosylhomocysteinase	3.3.1.1
adenosylmethionine decarboxylase	4.1.1.50
adenosylmethionine hydrolase	3.3.1.2
adenylate cyclase	4.6.1.1
adenylosuccinate lyase	4.3.2.2
adenylosuccinate synthase	6.3.4.4
alanine aminotransferase	2.6.1.2
alcohol dehydrogenase (NAD^+)	1.1.1.1
alcohol dehydrogenase ($NADP^+$)	1.1.1.2
4-aminobutyrate aminotransferase	2.6.1.19
2-amino-4-hydroxy-6-hydroxymethyl- dihydropteridine pyrophosphokinase	2.7.6.3
AMP deaminase	3.5.4.6
α-amylase	3.2.1.1
arginase	3.5.3.1
aspartate aminotransferase	2.6.1.1
aspartate carbamoyltransferase	2.1.3.2
ATP synthase (H^+-transporting)	3.6.1.34
Ca^{2+}-transporting ATPase	3.6.1.38
cAMP-phosphodiesterase	3.1.4.17
carbamoyl-phosphate synthase	6.3.5.5
carbonic anhydrase	4.2.1.1
catalase	1.11.1.6
choline acetyltransferase	2.3.1.6
chymotrypsin	3.4.21.1
CoA transferase (succinate/acetate)	2.8.3.8
cytochrome oxidase	1.9.3.1
cytidine deaminase	3.5.4.3
cytosine deaminase	3.5.4.1
dihydrofolate reductase	1.5.1.3
dihydrofolate synthase	6.3.2.12
dihydrolipoamide acetyltransferase	2.3.1.12
dihydrolipoamide dehydrogenase	1.8.1.4
dihydroneopterin aldolase	4.1.2.25

ribonucleoside-diphosphate reductase	1.17.4.1
serine hydratase	4.2.1.13
spermidine synthase	2.5.1.16
spermine synthase	2.5.1.22
succinate dehydrogenase	1.3.99.1
succinate semialdehyde dehydrogenase	1.2.1.24
succinyl-CoA synthetase	6.2.1.4
superoxide dismutase	1.15.1.1
threonine dehydrogenase	1.1.1.103
thymidine kinase	2.7.1.21
thymidine phosphorylase	2.4.2.4
thymidylate synthetase	2.1.1.45
triose-phosphate isomerase	5.3.1.1
trypsin	3.4.21.4
uracil phosphoribosyltransferase	2.4.2.9
uridine kinase	2.7.1.48
uridine nucleosidase	3.2.2.3
uridine phosphorylase	2.4.2.3
xanthine phosphoribosyltransferase	2.4.2.22

1
The Nature of Parasite Adaptation

1.1 Introduction – what is a parasite?

P is for *parasite*.

Parasites are punctual, prudent, prolific, periparturient, pampered, pandemic, particular, pathogenic, peculiar, pecunial, pejorative, pleiomorphic, penetrating, peragrate, perforating, periclitate, perineal, peripatetic, permeable, pernicious, perpetual, perplexing, persevering, pertinaceous, perturbatory, pestilential, piercing, pleonastic, poisonous, polygametic, populous, polyphyletic, practical, prevalent, problematic, profligate, progenitive, promiscuous, prompt, propinquent, protean, proximate, provident, palpable, punctilious, pyrogenic.

Parasites are <u>not</u> parsimonious, pathetic, peaceful, penitent, pessimistic, perambulatory, pleasing, poetic, pretentious, pretty.

(With apologies to H. McL. Gordon).

The possible interactions of one species of organism with another form a continuum. Associations range from the condition in which both organisms are benefited to that in which both are damaged. In this continuum, parasitism occupies a sort of middle ground.

When an association occurs in which both species are benefited it is known as *symbiosis*. There are many examples of symbiosis, some of them so well known that they are frequently encountered in primary and secondary school text-books. They include sea anemones that are transported by hermit crabs, the one gaining a supply of food, the other, protection; tuatara lizards and petrels living in the same burrows; and the more intimate associations of algae and bivalve molluscs or of algae and fungi to make the composite lichens. An even more esoteric example is the eukaryotic cell itself which, it is now widely believed, is "a product of ancient symbioses" (Margulies 1981). The eukaryotic cell involves at least three and possibly four symbionts (the 'host' cell, chloroplasts, mitochondria and possibly the microfibrillar structures involved in the spindle, centrioles and basal granules of cilia and flagella). And at least one example exists today of a 'pentad', a symbiotic association of five organisms. *Mixotricha*, itself a symbiont of an Australian termite (which thus becomes a hexad!), comprises the 'host' cell, three sorts of cortical bacteria and one cytosolic bacterium. In many cases the interdependence between symbionts is so great they they cannot survive outside the association.

However, symbiosis is a 'portmanteau' word that covers a variety of associations. At one level, that of *inquilinism* , only one of the pair of associating organisms benefits, the other being unaffected. A strict definition of an inquiline is an animal which lives in the nest or abode of another; a commensal or, according to the Shorter Oxford English Dictionary (1973), a guest, but it can be expanded to include a number of other relationships. For example, it may be used to describe the inadvertent transport of a small and insignificant species by a larger one. The mite *Rhinoseius* is a nectar thief that lives in flowers visited by humming birds and hitches rides on the birds' bills (Colwell 1973). The mite benefits by improved dispersal while the bird is probably not even aware that it has been so used. A similar phenomenon sometimes occurs *within* a species, when juvenile stages are transported by adults. This process, known as phoresy, increases the dispersion of the juvenile stage but also improves the reproductive success of the adult, and therefore, strictly, falls outside the definition.

Inquilinism blurs into *commensalism* which means, literally, 'at the same table'. The associating organisms utilise the same resource but only one benefits, the other or others being neither advantaged nor disadvantaged in the process. Much of the intestinal flora of mammals can be regarded as commensal, neither harming nor helping the host but dependent on it for warmth and protection. Of course, many organisms in the gut are beneficial to the host – the lumen bacteria, for example – and commensalism thus shades into *mutualism*. In mutualism, both or all partners benefit, as in the case of the cleaner fish that remove parasites from larger fish.

Amensalism occurs when one organism is disadvantaged simply by the presence of the other. The case of buffalo trampling frogs and flowers while drinking at a water-hole falls into this category; no one will deny that the amphibians and plants are disadvantaged but, no doubt, the buffalo are supremely indifferent to their presence! Next, in this continuum of associations, one might include *competitive exclusion* . This is a formal definition of Gause's principle, a truism that states that no two organisms may occupy the same ecological niche. In one of the manifestations of competitive exclusion, both species may be disadvantaged as they compete for a limited resource.

Finally, there is *parasitism*, in which one organism suffers to the advantage of the second. These relationships can be neatly summarised by the matrix in Figure 1.1.

SPECIES 2

	0	+ –	–
0	00	0 +	0 –
+	+ 0	+ +	+ –
–	– 0	– +	– –

SPECIES 1 (row label, at left of second/third/fourth rows: 0, +, –)

0 implies that the species concerned gains or loses nothing by the association; +, that it gains and -, that it loses. In the matrix, the condition for species 1 is entered first, followed by the condition for species 2. This leads to six types of interaction, expanded to nine by a simple reversal of polarity. They are: 00, no interaction; +0, inquilinism or commensalism; ++, mutualism; -0, amensalism; -+, parasitism; --, competitive exclusion.

Figure 1.1: A Two-way Matrix Showing the Possible Interactions of Two Species

Unfortunately, the definition of parasitism given above is really so general as to be almost worthless. For example, it would almost certainly include many types of predation, although it might exclude those well balanced prey-predator relationships where consistent culling of unfit individuals maintains the inclusive fitness of the group of prey animals. Price (1980), in his study of the evolutionary biology of parasites, avoids the question by falling back on a dictionary definition:

> "a parasite is an organism living in or on another living organism, obtaining from it part or all of its organic nutriment, commonly exhibiting some degree of adaptive structural modification, and causing some degree of real damage to its host" (Webster's Third International Dictionary).

This is also a very broad definition and is adopted by Price because it suits his thesis – no great sin, this, as we are about to do a similar thing! It permits him to include within the definition of parasite those insects that may spend the whole of their lives feeding on a single tree – not because they could not move to another tree but because a tree represents such an enormous resource to a small organism like an insect that it is unlikely to exhaust it. A more usual category for such insects would be 'pest'!

At the other end of the scale is MacInnis' 1976 definition of parasitism. ".. the case in which one partner, the parasite, of a pair of interacting species is dependent upon a minimum of one gene or its product from the other interacting species, the host, for survival." This seems to be a precise statement about the relationship between host and parasite but closer inspection shows that it leaves the field as wide open as the definition in Webster's Dictionary. For example, humans need a dietary source of ascorbic acid and 'interact' with citrus fruits. Ascorbic acid is undoubtedly produced by lemon trees as a result of the activity of several of their genes and gene products. Few would defend the position that humans parasitise lemon trees, or lettuce, or the lactobacillus that produces yogurt. These activities might reasonably be called browsing or grazing. This slightly ridiculous example does, however, serve to point up the difficulties attendant on the search for a definition of parasitism. All animals interact, at a trophic level, with other organisms and are therefore dependent on their gene products. Even the introduction of the concept of specificity towards a 'host' is no help as some predators are exquisite specialists while many parasites are positively promiscuous in their choice of host.

Price (1980) does, however, introduce a new element into the Webster's Dictionary definition. He emphasises the point that an individual of any parasitic species will usually gain the majority of its food from a single living organism. He adds that "although a species of parasite may utilise several or many host species, each individual obtains most of its nutrition from an individual host. Species with complex life cycles may exploit two or three hosts in a predictable sequence". While this is still not a tidy statement and requires mental reservations and glosses, it provides us with a working concept.

The search for a universally acceptable definition has occupied many people for many years and it would foolish of us to imagine that this book could make any contribution to those arguments. Instead, we feel that it is much more useful to define the limits of the book and attempt to arrive at a simple functional definition. The following chapters, then, are concerned with parasitic Protozoa and helminths. Ectoparasites are excluded, not because they are not important but because relatively little is known about their adaptive biochemistry. By restricting discussion to endoparasites, we can therefore adopt the useful but limiting statement that a parasite is an organism that lives within another organism, called a host, to that host's detriment, and is absolutely dependent upon that host for the completion of at least a part of its life cycle.

1.2 The host-parasite relationship

The host-parasite relationship is the most intimate of all relationships and, because of this, is extremely difficult to study. The environment of a parasite is the inside of another organism. A parasite removed from its host is an organism *in extremis*, severed from its life support systems. The host provides a physical and chemical *milieu*, a space with important characteristics of pH, oxidation-reduction potential (redox) and availability of nutrients and inorganic compounds such as carbon dioxide. It also offers a complex biochemical environment in which intermediates and hormone regulators of host metabolism create a background of biochemical noise against which the parasite must complete its own life cycle. Under these circumstances, it is not surprising that the cues that initiate the development of each parasitic stage are relatively simple, loud, unambiguous 'shouts' that can be 'heard' by the parasite above the background of intense activity that is the host. These stimuli, or developmental cues, often exert the most subtle influences on the life of the parasite. For example, protoscoleces of several strains of the hydatid organism, *Echinococcus granulosus*, can be cultivated in nutrient media in the laboratory. In one strain contact with the right kind of surface is the cue that determines whether development will proceed in the direction of segmentation and elongation – as in the gut of the definitive host, the dog – or in the direction of cyst formation – as in the tissues of an intermediate host such as the sheep (Smyth 1969).

Parasites, then, are only one half of a host parasite relationship. The other half, the host, continually reacts to the parasite at all levels of organisation. The environment occupied by parasites is not very hospitable. Because a parasite is 'not self' to a host, it either has to hide from the recognition systems with which the host is equipped to defend itself, or it has to overcome them by defences of its own. The recognition systems are, of course, the immune responses of the host.

In higher organisms, mechanisms of immunity are complex (see Chapter 4) but, broadly, they can be separated into three sorts of components, mechanical, cellular and humoral. Mechanical ones are simple physical barriers that prevent a parasite establishing in its site of predilection and completing its life cycle. Heavily keratinised skin may be sufficient to prevent the entry of delicate larval forms into hosts or prevent penetration by the mouthparts of an insect vector.

The remaining two types of response, cellular and humoral, depend on the existence of 'recognition molecules'. The classical example is the antigen-antibody system. Antigens are molecules produced by the parasite which, being 'not self', are recognised by and elicit a response from the host. It is therefore advantageous to the parasite to expose as few of these antigenic molecules to the host as possible. In this way, any adverse effects that the host may have on it are minimised. Some, such as the malaria parasite, possess decoy antigens. These are molecules of great antigenicity but of little importance to the parasite except in their capacity to elicit a massive and largely irrelevant immune response from the host (Sharma, Svec, Mitchell and Godson 1985). Other parasites exhibit antigenic mimicry (Damien 1964). They coat themselves with molecules so like host molecules that they are not recognised as 'not self' and develop undisturbed. Some schistosomes apparently do this (Clegg, Smithers and Terry 1971). Still others, such as the trypanosomes, regularly change their surface antigens (Vickerman 1978). Thus, although the host may be alerted to the presence of the parasites and mount a response, the parasites change their surface antigens under its influence so that the response becomes inappropriate.

Other parasites actually suppress the immune response of the host, rendering it less effective and increasing the chances that the host will be invaded by other unrelated organisms. The mechanisms of immunosuppression are many and not clearly understood. At present, they seem to fall into two categories. The first is mechanical, in which the components of the immune system are prevented in some way from having access to the parasite, as in the case of mice infected with *Mesocestoides corti* (Mitchell and Handman 1977). The second is humoral, with the parasite secreting a substance that interferes with the sequence of events that gives rise to an immune response. An excellent example of this is *Leishmania*, whose successful invasion of the macrophage depends on inhibition of the complement cascade (Blackwell, McMahon-Pratt and Shaw 1986).

1.3 The evolution of the host-parasite relationship

Parasites do not evolve in isolation. The host-parasite relationship evolves and the changes that take place in the host are just as important as the changes that take place in the parasite. Evolutionary changes in the host-parasite relationship are not necessarily changes in form; they may also be changes in function. Form is conservative, and taxonomists often use the similarities between parasites to establish relatedness in vastly dissimilar groups of animals. This is not surprising when one considers that the shape of an animal is a function of the mechanical stresses it experiences in its environment. Thus, for metazoan parasites that live in intestines, elongate, fusiform or ribbon-like shapes are highly probable. Evolutionary changes in the morphology of organisms are obviously relatively easy to identify. Changes in function may evolve independently of form, and are much less easy to detect.

Coevolution, the parallel and interdependent evolution of host and parasite, is therefore a characteristic of the host-parasite relationship (Thompson 1982). Parasites usually have a shorter generation time than their hosts so that the relationship is unequal, or asymmetrical. Parasites are therefore pacemakers and hosts must run the Red Queen's race in an attempt to keep pace with their parasites. A major problem in understanding the host-parasite relationship lies in the inadequacy of our knowledge of the parasites and in our inability to ascribe unequivocally any of their biochemical functions to the parasitic habit. As Fairbairn (1970) has pointed out most of the biochemical peculiarities of parasitic helminths can, with equal logic, be ascribed to epigenesis, the interaction of gene products to produce the phenotype. Our inability to identify accurately particular characteristics as adaptations to either the parasitic habit or to epigenesis does not, however, preclude the use of the concept of coevolution to assist our understanding of the processes of biochemical adaptation.

An implication of Fairbairn's *caveat* is that many of the biochemical 'adaptations' that are generally regarded as typical of parasites are really adaptations to certain physical and chemical properties of the environments often encountered within hosts, and are only incidentally to do with parasitism. According to this view, the host intestine is, for example, simply a special case of a class of environments that have, among other attributes, a pH of around 7, low redox potential, low availability of oxygen and high carbon dioxide concentrations. These features are encountered in other biomes, for example, anoxic mud. The most notable of these biomes is the sulphide layer. The sulphide layer is the black, reduced sediment, often several metres thick, that is found underneath the yellow, oxidised sands that form the beds of oceans and the floors of estuaries and lakes. It is enormous in area and volume – after all, more than three-quarters of the planet is covered by water – and of immense antiquity, having been in continuous existence since the first appearance of oxygen in the earth's atmosphere. This biome is occupied by a variety of ancient Protozoa and Metazoa that have biochemical features similar to those of many modern intestinal helminths. It seems probable that possession of these biochemical attributes was important for organisms that assumed the parasitic habit but they cannot be regarded as a consequence of that habit (Bryant 1982).

Unlike the metabolic pathways found in the higher Metazoa, those of the Protozoa, Platyhelminthes and Aschelminthes show considerable variation *between* groups. Further, the pathways of energy metabolism in rats and mice or mice and men are very similar. There are, however, several different modes *within* a single taxonomic group of parasites. As Chapter 6 shows, metabolic pathways, and hence metabolic functions, vary within species of parasite and are mutable even on a short timescale of less than a decade. We do not know to what extent these variations are 'adaptive'.

1.4   The role of parasites in evolution

It has been conventional to think of the host-parasite relationship as one that, after a stormy start, sails into the peaceful harbour of symbiosis. If this were true, then it follows that those parasites that are the most pathogenic are also the most recently established. Pathogenicity could thus be taken as a sign of poor adaptation. Reasoning superficially, it seems to be of no immediate advantage to a parasite to kill its host, and once this doubtful point is accepted, an inevitable consequence of the coadaptation of host and parasite must be decreased pathogenicity. The immune response of the educated host keeps its parasite burden low but does not achieve a sterilising immunity. At the same time, the educated parasite maintains its numbers at the maximum compatible with minimal damage to its host.

It is now clear that this scenario is only one of a range of possible scenarios

(Figure 1.2). For example, in parasites that require the death of their host for transmission to the next host, rapid reproduction and increased pathogenicity, which would cause the death of the host more quickly, may give an evolutionary advantage to that particular population.   Alternatively, the host's death may be a by-product of the processes required to maintain large numbers of parasites accessible to a vector.

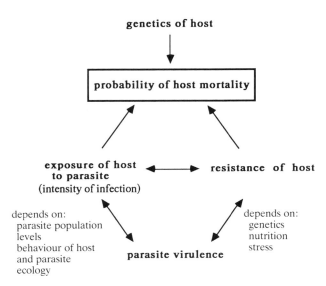

Interacting characters are indicated by double headed arrows; those that directly affect mortality are indicated by single arrows.

Figure 1.2: Characteristics that Determine the Outcome of Parasitic Infections (Modified after Holmes and Price, 1986)

Ewald (1983) has argued that the facile assumption that all host-parasite relationships tend towards a benign equilibrium, with its corollary that the most recently established relationships are likely to be more pathogenic for the host, is incorrect.   The establishment of a benign relationship and, ultimately, true symbiosis is the result of only one of several possible evolutionary trajectories. Another possibility is evolution to a moderately pathological relationship, as in the case of vivax malaria. The recurrent febrile periods induced by the parasite may make the sufferer more attractive to mosquitoes and help to ensure the transmission of the disease. Finally, it may be in the interests of the parasite to cause debility or the death of its host. This is particularly true for parasites that inhabit muscle and other tissues of their intermediate hosts. Debility may make an intermediate host such as a caribou more susceptible to predation by a carnivorous definitive host such as a wolf. In the absence of the assistance of an active predator, the last resort of the parasite might be to cause death of the intermediate host so that a scavenger may become infected. Alternatively death may ensue because the immune response of the host induced an excessive and inappropriate pathological response.

A number of consequences follow from the concept of evolution towards increased pathogenicity. If the definition of parasitism is stretched to cover such animals as sap-sucking aphids (which could be considered a form of ectoparasitism), it is probable that up to half of all known species of animals are parasitic. This idea is the foundation for a recent ecological hypothesis (Hamilton 1980) that suggests that interactions *between* biota are more important in driving evolution than interactions between biota and environment. The impact of parasites on their hosts is large because of the differences in cycling frequency between host and parasite life cycles. Parasite species are capable of passing through more generations in a given time than their hosts. If the life cycle of the parasite includes regular genetic recombination, its opportunities for variation and adaptation are greater than those of the host in the same time interval. Thus, host and parasite coevolve, but the more rapid evolution of the parasite could be a significant factor influencing the evolution of the host. A good example is the proposed role of malaria in maintaining otherwise deleterious haemoglobin variants in human populations. Thalassaemia and sickle cell anaemia cause changes to red blood cells that render them inhospitable to the invading parasite.

The co-evolution hypothesis has been taken to greater lengths by one school of thought (Zuk 1984) that argues that the evolution of sex derives from the impact of the parasite burden. The argument depends on the idea that sex provides the host with sufficient genotypic variability to allow it to adapt to parasitic attack. The same argument applies, of course, to the parasites themselves, which makes this hypothesis somewhat circular. Did parasitism evolve before sex (i.e. before genetic recombination)? If so, it is difficult to envisage how organisms without genetic recombination could have coped with the challenges and pressures of environmental variation. Then, since most parasite species undergo genetic recombination during their life cycles, does this mean that parasites themselves had parasites? Is it not more likely that they evolved from free-living organisms that had sex, and possibly parasites? If the answer is 'no', are the original, non-sexual, parasites extinct, or are there still some primitive ones extant? It can be argued that the genetic variability required to be a successful parasite – at least early in evolution – would have required genetic recombination in these groups too.

1.5 The frequency of parasitism

It is, perhaps, fanciful to claim that the normal condition of life is the parasitic habit but it is not stretching truth too far to claim that parasites make up an enormous part of evolutionary diversity. If one makes the assumption that every organism has at least one specific parasite then we must conclude that at least half of the living world is parasitic (if one excludes considerations of biomass and counts only species). This is, of course, a maximum estimate, for many parasitic organisms are promiscuous with respect to their hosts, spreading their favours across a wide and diverse range of organisms. So, if parasitism is not the normal habit of life, it is at least a very common one.

In support of this view, and in accordance with the definition of parasitism given above, Price (1980) has compiled a very interesting table in which he classifies British insects on the basis of their feeding habits. The table shows that 35.6% of the insect fauna are parasitic on animals, 35.1% are parasitic on plants, 21.5% are saprophagic, living on decaying organic detritus, 3.9% are predators and 3.8% are non-parasitic herbivores and carnivores. The number of species is given as 16,929. Of those listed as parasitic on animals, the Hymenoptera comprise by far the greatest number of species while the Lepidoptera are considered to be major parasites of plants. Many would regard the Hymenoptera as predators rather than true parasites while few would admit that Lepidoptera are parasites. However, even with these reservations, and excluding both the Hymenoptera and Lepidoptera from consideration, the proportion of parasites is almost 26%. It is obvious that the basic thesis that parasitism has a very high incidence can be sustained, for insects at least. The numbers of parasitic species will, of course, vary from class to class and phylum to phylum. Some phyla, like the Chordata, possess few species that are parasitic, while others, like the Platyhelminthes, contain large numbers.

1.6 Ecological aspects of parasitism

The 'world view' of a parasite must be similar to that of the Bedouin, for whom life was once a series of dangerous journeys across large areas of extremely inhospitable desert to tiny, welcoming oases that harboured all the necessities of life. An oasis is a resource to be exploited. At the risk of overburdening the metaphor, both the parasites and the Bedouin are adapted to extremely patchy environments. The desert of the parasite is the environment in which its host-oasis dwells, whether forest, savannah or ocean. Potential host organisms are separated in space by a parasitological desert and potential host populations are widely dispersed in this hostile environment. Transmission – getting from one patch of resources to another – is the paramount problem.

It would be a truism to observe that the success of a parasite depends on how well it copes with the problems of transmission. Obviously, all extant parasites have coped with it successfully while extinct ones have failed in this or some other department. Relative success can be measured as the proportion of a host species that carries the parasite. It is, of course, only one measure, as factors such as host density may have an over-riding effect, but there are some interesting observations that may be made when dissimilar parasites exploit the same host. 'Successful' parasites would include the malarias (that together infect some 25% of the human race) and the threadworm, *Enterobius vermicularis*, that probably has an incidence of about 10%. Each has solved the problem of transmission in a different way. Transmission of malaria involves a vector, the mosquito, while that of *Enterobius* depends on anal-oral transfer. Malaria harnesses the biting habit of its intermediate host to ensure transmission while *Enterobius* exploits the behaviour of its definitive host.

These two examples, then, show that the problem of patchiness of resources and their successful exploitation can be solved by widely divergent parasite strategies. These strategies include a marked capacity for sexual and asexual reproduction. Reproduction is achieved either by the production of spores or eggs, or by the interpolation of a developmental stage that facilitates transfer to another host and may even offer the possibility of another proliferative stage. The capacity for unchecked multiplication is an important characteristic of all organisms that exploit patchy environments; it is an opportunistic and efficient strategy for the rapid utilisation of a resource. The Platyhelminthes have this characteristic developed to a very high degree.

From the point of view of the parasite, hosts are separated in not three but four dimensions. They are separated in space but they are also separated in time. Very often passive means are adopted to overcome the two barriers. Thus, some nematodes are equipped with resistant eggs that can withstand wide variations in climatic conditions for considerable periods of time. The embryos are inert, undergoing little development until ingested by another host animal. Alternatively, the eggs may hatch into larvae, which possess certain behavioural characteristics that improve their chances of colonising another host. Hookworms have aggressive larvae that burrow into their definitive hosts. Strongyles, such as *Haemonchus contortus*, possess larvae with behavioural traits that position them on accessible pasture at the right time of day to improve their chances of being eaten by grazing sheep.

Such transmission, however, relies on high population density of the host animal and its more or less continuous occupation of a geographical area. These criteria are fulfilled in the case of intensive farming of domestic animals, when the population of gastro-intestinal nematodes builds up to intensities that threaten the life of stock. In the wild, the population densities of potential hosts may not be high. They may be so low that the probability of the eggs – or other infective stages – generated by the parasites from one host encountering a second host is vanishingly small, during the time that they are capable of living free. Under these circumstances a second, intermediate host may provide the bridge in time and space between one definitive host and another. The distribution of the intermediate host must overlap that of the definitive host, its lifespan must exceed the critical period between successive hosts and its population density must be sufficiently high to compensate for the low population density of the definitive host. All of these factors put a very heavy selection pressure on parasites and account on the one hand for the fecundity of parasitic organisms and, on the other (as a consequence of fecundity), for their variability. In extreme circumstances, variants with a high rate of survivorship are selected and may even become fixed in the population, provided the particular selection pressure that promoted them is sustained. Thus a new strain of parasite is established.

1.7 What is a strain?

Biologists have always been concerned with taxonomy. A uniform system of classification of organisms is an essential prerequisite for any sensible communication about them, so it is unfortunate that the proponents of different taxonomic systems have for many years rendered the scientific literature more interesting but less helpful by their advocacy. Animal taxonomy, however, now has a healthy antiquity and we no longer have to worry whether coelenterates and molluscs belong to the old class Vermes. There is now considerable agreement about the taxonomic status of the higher taxa, although the means by which it is established – phenetically or cladistically – still generates much heat in the pages of the taxonomic journals.

Although there is generally little dispute about the definition of a genus of organisms, and it is accepted that a genus comprises a cluster of related species, there is still argument about what exactly constitutes a species. It is out of place here to probe too deeply into the discussion about the multitude of definitions. It is sufficient simply to point out that the most useful definition is a practical one, and the most practical one is that 'species are reproductively isolated populations'. They *do not* normally interbreed. The expression 'do not' is used advisedly; 'cannot' would imply a degree of separation not even found between lions and tigers. They do not exchange genes with neighbouring populations because of geographical, climatic, temporal, ecological, behavioural or physiological barriers. Gene flow is thus circumscribed by the boundaries of the population. Quite clearly this definition of species can cause trouble. The tapeworm *Hymenolepis diminuta* is found in rats in America, Europe and Australia. Can they be considered to be three separate species? They do not interbreed (but would if given the chance), and at least three separate isolates from the rats of those continents have been shown to have different developmental, physiological and biochemical characteristics (Bryant and Flockhart 1986). It is obviously necessary to use a little common sense and choose definitions according to the use to which they will be put. The criterion of reproductive isolation is of little use to the systematist and taxonomist, who is concerned with introducing order into an apparently disordered world. But it opens the way to the insights of the evolutionary biologist.

The separation of a population into two or more isolated, non-breeding sub-populations is an end-point of the process by which reproductive isolation occurs. The process is usually a continuous one. Adjacent populations can exist between which there are various degrees of gene exchange. If high levels of gene exchange are maintained, it is unlikely that the two populations will diverge sufficiently to be counted as new species. At very low levels of gene exchange it is very probable that divergence into daughter species will occur. Somewhere in this continuum of gene flow lies that critical point, different for each emerging species, at which a speciation event can be said to have taken place.

Strain variation is found in divergent populations before that critical point is reached. For obvious practical reasons, biologists usually study populations, or small samples of populations, rather than the whole species. The biochemical literature on parasites contains many examples of apparent disagreement which can, in the end, only be explained by proposing the existence of distinct, in some cases laboratory-adapted, strains of parasites. This can cause problems. Parasite biochemists, like all scientists, strive for generalisations, hoping that what is true for one protozoan or helminth will be true for related genera, classes, or even phyla. It will be clear from subsequent chapters that, regrettably, this is not so. We have found, in fact, that generalisation even within a species is not always possible, because of intraspecific variation in biochemical attributes.

How do we recognise a strain? Unfortunately, its definition depends on authority. This was clearly recognised by Charles Darwin (1859) in the *Origin of Species*:

> "No one has drawn any clear distinction between individual differences and slight varieties; or between more plainly marked varieties and subspecies, and species ... what a multitude of forms exist, which some experienced naturalists rank as varieties, others as geographical races or sub-species and others as distinct, though closely allied species."

Later, in the same argument, he comments that species are only marked and permanent varieties. For the purposes of this discussion, the term 'strain' is assumed to include the terms 'variety', 'race' and 'sub-species'.

The criteria for establishing the existence of a strain are many. The classical taxonomist uses a complete range of morphological data, such as linear dimensions, ratios of measurements, relative positions of internal and external organs, and numbers of repeating parts such as bristles, hooks or segments. There is no problem accepting these apparently objective criteria because they are easily accessible with lens and graticule. But this level of taxonomy has proved no longer suitable for many applications where non-morphological differences are apparent and must be categorised in a useful way. Classification has become more difficult. For example, a change in the architecture of a mitochondrion, seen under the electron microscope, is a morphological change that probably reflects a more fundamental physiological or biochemical change. Alternatively, a morphological change in a mitochondrion may be the *cause* of a fundamental biochemical or physiological change. Whatever the case, such a change is not immediately evident to the eye and requires more extensive investigation to be brought to light. It must then be shown that this change is reliably inherited and causes its possessors to be different in some significant way from the rest of the species. The question of what is 'significant' and what is 'trivial' is important and at present seems to depend on the techniques used to detect the differences in the first place.

In the end, therefore, judgements depend on the reliability of the technique used to acquire the information and the consistency with which a particular character is detected. An important element is the subjective decision about how great the variations have to be before they change from 'trivial' to 'significant'. Taxonomists have been slow to utilise the techniques of biochemistry, largely because they may be technically complex and expensive. The two major biochemical techniques that have recently come into use by taxonomists are enzyme separation procedures using electrophoresis or isoelectric focusing, and DNA comparisons using either physical properties of DNA such as buoyant density, hybridisation studies, or mapping using endonuclease digestion. The application of such studies is discussed in more detail below. First, we will discuss the significance of strain variation.

1.8 Are strain variations adaptive?

It is not at all clear whether strain variations are adaptive. Since the strains in question are found surviving and flourishing, each in its particular niche, we must conclude that they are appropriately adapted for such environments. We may indeed have observed strains arising as a result of a definable environmental change, and been able to observe speciation events taking place. But, there is considerable randomness in the process of establishing a strain; variants are produced all the time and are the raw material upon which selection works. The types of variants available at the extremes of the population may differ, so that their response to a given type of selection pressure may differ also. Thus, different strains may emerge at the extremes of the range even though they are apparently responding to the same selection pressure. In each case the variant is 'adapted' to the same environmental change, but the variants are different from each other. Which variant is truly 'adapted'?

This problem is further compounded by the question of whether parasites – or any organisms, for that matter – are precisely adapted. There has been considerable debate about this in recent years. The 'adaptationist' would argue that every feature of an organism is a product of natural selection and is precisely adapted for some function. Sometimes these functions are obscure, so that the observer admits to ignorance; at other times, a hypothetical function is proposed. If subsequent observation invalidates the hypothesis, a new one is generated. Adaptationists do not abandon the 'adaptationist' program.

An alternative view, which Gould and Lewontin (1979) have advocated, using the unlikely analogy of cathedral architecture, is that the evolution of an organism and its development are compromises with what is possible. In the cathedral example, a 'spandrel' is the space between the curvature of the arches on a supporting pillar. The pillar and the arches are called 'adaptive' because they have the important function of supporting the roof. The spandrel is simply an area of stonework that is present because it is required by the geometry of the structure. The architects (evolution) may subsequently decorate (modify) the spandrels to improve their appearance. Thus it might appear to the observer that the primary function of spandrels is not support but decoration. Hence the origin of the spandrels has been masked by the subsequent decoration. (Of course, this argument should not be carried too far: decoration of the spandrels was presumably 'adaptive' in that it served a human social function – such considerations do not normally enter into this type of evolutionary discussion!)

There must be many 'spandrels' in the architecture of organisms. The final form of an organism and the regulation of its functions at the physiological and biochemical levels have to sacrifice adaptive precision to the physical constraints of geometry, or to the chemical ones of molecular interaction and the maintenance of genetic variability.

Truisms are useful for drawing attention to apparently self-evident facts that are easily overlooked, and here is an important one. Organisms are adapted at all stages of their life cycles. Helminth parasites, like insects, have a number of discrete stages in their life histories. Each stage must carry a genetic program to enable it to survive in the appropriate environment and to develop to the next stage. Further, the adaptation of a larval form to one particular environment is unlikely to be wholly independent of the adaptation of the other stages. Larval adaptation is constrained to some extent by the requirements of the adult and *vice versa* . Thus an organism is subject to a variety of selection pressures, which may act in different ways at different times. The final form will be a compromise between 'perfect' adaptation and what is possible, given the particular starting materials and a particular set of environments. In fact, the compromise itself is the 'perfect' adaptation, because it *retains flexibility* . Darwin, as Ernst Mayr (1983) comments, recognised this:

> "Evolutionary change falls far short of being a perfect
> optimisation process. Stochastic processes and other
> constraints upon selection prevent the achievement of
> perfect adaptedness. Evolutionists must pay more
> attention to these constraints than they have in the
> past. However, as already stressed by Darwin (1859,
> p. 201), there is no selective premium on perfect
> adaptation."

At the biochemical level, it is not possible to say that anything unusual or slightly different observed in a parasite is necessarily an adaptation to the parasitic habit. For example, consider the anaerobic respiratory pathways that utilise the fumarate reductase system. The cellular *milieu* within the host that makes these pathways adaptive is characterised by low oxygen tension, and high carbon dioxide levels. These characteristics are also encountered within the closed valves of molluscs at low tide, in anaerobic mud and in the 'sulphide system' (see Section 1.3). Molluscs and the free-living organisms in anaerobic muds have biochemical characteristics that would enable them to survive in the *physical* environment of the vertebrate intestine. Permanent establishment as parasites, however, depends on finding solutions to the other problems posed by isolation, host immune processes, digestive mechanisms, and competition from the existing denizens of the gut.

'Strain' variation in parasites is clearly an important phenomenon that deserves the close attention of parasitologists because it is the basis of evolution. There are practical reasons, too. Parasites vary in pathogenicity for their hosts and in resistance to the treatments used to combat them. It is essential in medical and veterinary medicine to be able to recognise the variants, understand what causes them and to appreciate their potential for further variation.

1.9 Biochemical adaptation and life cycles of parasites

Complex life cycles pose special problems for the regulation of development. Unfortunately, little is known about the genetic constitution of parasites but a few generalisations are possible. First, parasites, like all organisms, are adapted at all stages of their life cycles. The corollary of this statement is that natural selection acts on all life stages, so that the parasite must co-evolve with several hosts. Many parasites resemble insects in that their life cycles are discontinuous and a different set of genes has to be activated for each stage of the life cycle. For example, the trematode *Dicrocoelium dendriticum* is successively parasitic in snails, ants and sheep. In each host, different combinations of genes must be activated and regulated to provide the parasite with the necessary structural and functional requirements for survival. Each stage requires a specific set of genes to be activated, although it is highly likely that there is a basic suite of 'housekeeping' genes whose activity is common to all stages. For example, the pathway of glycolysis is an ancient series of reactions that is almost ubiquitous in pro- and eukaryotic cells. Presumably, this metabolic sequence is functional in all stages of the life history of *Dicrocoelium*. On the other hand, there is much evidence to show that the respiratory pathways in the mitochondria of parasites, the aerobic and anaerobic electron transport chains, are switched on and off according to the developmental stage and the exigencies of the environment in which the parasite finds itself. Figure 1.3 shows that embryonated *Ascaris* eggs possess an aerobic electron transport system with a classical, mammalian type of terminal cytochrome oxidase; in the adult there is much greater emphasis on an alternative

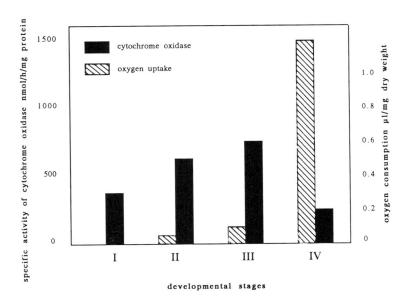

The stages of development are as follows: I - 1 to16 cells; II - 32 cells to gastrula; III - gastrula to first stage larva; IV - second stage larva. Data are averages for each stage, as reported by Oya, Costello and Smith, 1963

Figure 1.3: Changes in Cytochrome Oxidase Activity and Oxygen Uptake during Development in *Ascaris*

oxidase (Oya, Costello and Smith 1963). The development of *Fasciola hepatica* also demonstrates this point (Figure 1.4). Young liver flukes possess an active tricarboxylic acid cycle, but by 24 days post-infection it is subordinate to aerobic acetate production. Later still, at 114 days, these pathways are barely detectable and the major energy-producing metabolic sequence is anaerobic.

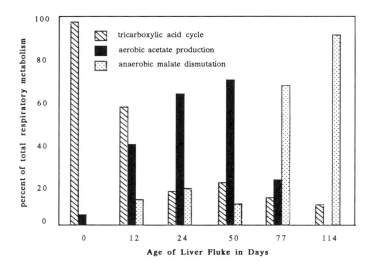

Figure 1.4: Percentage Contribution of the Three Pathways of Glucose Catabolism during Aerobic Incubation of the Liver Fluke, *Fasciola hepatica.* (Data from Tielens, Van den Heuvel and Van den Bergh 1984)

Another means of ensuring transmission is for the parasite to affect one host in such a way as to improve its chances of encountering the next one. The classic example of this is *Dicrocoelium*, a larval stage of which invades the nervous system of the ant. The effect of parasitism is to change the behaviour of the ant so that it migrates to the tips of grass stems, increasing the likelihood that it will be eaten by a grazing animal. There are, however, even more subtle ways of achieving the necessary result. One such way is to induce a pathological response in the host.

1.10 The physicochemical characteristics of parasite environments

"Sometimes the maggots died when the nut was cracked. The story had it that the poor little maggots died of shock. They believed that the interior of their nut was their whole world, and the wrinkled case that contained it the sky. Then one day, their world was cracked open. They saw with horror that there was a gigantic world beyond their world, more important and brighter in every way. It was too much for the maggots, and they expired at the revelation". (Aldiss 1982).

The psychic shock that Aldiss' maggots received when their world was cracked open is hardly less than the impact of the physiological shock that parasites receive when they are removed from their hosts into the glaring immensity of the laboratory. The parasites themselves make elaborate preparations for traversing the vastness between one host and another. Unless the experimenter makes similarly elaborate preparations for their cultivation they will quite quickly succumb to the absence of attachment sites, salt imbalance, to loss of their internal fluids, desiccation and the lack of suitable nutrients.

The cultivation of parasites in the laboratory is thus an extremely complicated affair. All parasites that have been successfully transferred to conditions *in vitro* require an aqueous medium buffered with salts to give the right pH and osmotic pressure; they require a source of respiratory energy, such as glucose and various amino acids, and other growth factors like the vitamins and their derivatives. They usually require a very precisely controlled gas phase, often a mixture of oxygen, nitrogen and carbon dioxide to provide exactly the right oxidising or reducing conditions and to support cellular respiration. Their environments have to be thermostatically controlled. They often have to be provided with just the right surface for attachment. When all this has been achieved, we may discover that our knowledge of the physiology of the parasite is incomplete and that some component of the medium is lacking. The usual way to overcome this is to add something derived from the host in the hope that it will remedy the deficiency. Sera of various sorts are used for this purpose, or bile, or other fluids, or even extracts or suspensions of cells. And finally, having done all this on the first day, we must set it all up again on the second day because, in the intervening period, the parasite has been excreting and secreting into the medium and rendering it uninhabitable.

All of these conditions, that we have laboured so hard to create, are, of course, naturally present in the normal environment occupied by the parasite. It resides in a space within the host where its requirements for pH, oxygen, carbon dioxide, and salts are automatically supplied and where the transport of nutrients to and waste products away from the parasite are taken care of by the circulatory mechanisms of the host.

1.11   Ectoparasitism and the invasion of the host

Any single animal provides many ecological niches in which a parasite might establish itself. At the simplest level it offers transport to places where there is greater nutrient availability or fewer predators. Alternatively, it may take advantage of its carrier's feeding habits and join in the meal, as does the pilot fish attached to a dolphin or a shark. From here, it is a short step to exploiting the surface to which it is attached, as do lampreys, as a source of food and (by burrowing under the skin) protection.

The space within a pipe is not part of the substance of the pipe; and an animal is really an elaborate pipe. The intestine of an animal is continuous with its external surface, so that an intestinal parasite is topologically (but not functionally) outside its host. Access to an intestine is via a number of cavities that provide ecological opportunities for potential parasites. The buccal cavity may be covered by a soft mucous membrane that is continually bathed in a fluid rich in organic nutrients. So too are the anus and rectum, and the reproductive and excretory systems. If the animal is a fish, another point of entry is through the gill clefts. Gills provide an excellent opportunity for a large number of parasites; a gill parasite is effectively internal, protected by the gill operculum, in a space with a very large area of expanded epithelium richly supplied with blood vessels, regularly ventilated by the activity of the fish.

Once in the pharynx, the potential parasite is well on its way to the lungs (if the host is terrestrial) or to the intestine. In the respiratory system, the parasite may occupy the trachea, the bronchi and bronchioles or the alveoli and lung parenchyma. In such environments it experiences high oxygen tensions. In the intestine, it may occupy the lumen, attach or burrow into the mucosa, or establish in a fermentation chamber, such as the colon of many mammals where it experiences a range of oxygen tensions down to zero. Alternatively, it may pass through the mucosa and migrate to other sites. In these intimate associations the rewards of parasitism are great but the penalties are proportionately large. Potential parasites of the intestine have to resist the digestive mechanisms, both physical and chemical, of the host; they have to withstand a more intensive immune response; and they have to cope with a changed physicochemical *milieu* engendered by the processes of digestion and the activities of the gut flora and fauna.

A number of diverticula branch from the vertebrate intestine; chief among them are the liver and the pancreas. These are environments readily accessible to parasites from the intestine, but they pose their own special problems to would-be colonisers. From the intestine, too, other tissues are readily invaded by parasites, either through lesions that have been caused by other means or through the ability of the parasite itself to burrow. Once into the peritoneal cavity, kidneys, reproductive organs, nervous, respiratory and vascular systems and skeletal musculature are all accessible, each site demanding its own special adaptations.

This scenario, describing the transition from ectoparasitism to endoparasitism, is vastly over-simplified. There are clearly many ways by which versatile parasites have achieved access to their sites of predilection. The scenario does, however, provide a sequence in which the constraints on parasites become more and more severe and the selection pressures increasingly intense.

At the surface of a large animal, the concentrations of oxygen and carbon dioxide are similar to those of the ambient environment. On hosts that live in an aqueous environment there is little change in surface water content. Terrestrial animals, however, confront the parasite with problems that are largely a result of the harsh, arid conditions on land. They have evolved epithelia or chitinous exoskeletons to limit the loss of water, electrolytes and other body constituents, and for mechanical protection. Such structures also prevent the entry of noxious molecules. There is, however, always some external secretion of important molecules available for the nutrition of external parasites. Further, in vertebrates there is cystine-rich protein available, as the skin continually sloughs and is renewed, together with various secretions that keep the skin supple and water-proofed. The keratinous coverings of scales, hair or feathers provide further ecological opportunities for insects like the lice. Other arthropods, such as the sarcoptic mites, actively penetrate into the *stratum corneum* while demodicid mites invade hair follicles. Biting insects and ticks probe and penetrate the dermis, causing local inflammation, exudation of lymph and bleeding, all of which allow them to feed. At this stage, the parasite comes into close contact with the immune system of the host, a fact that can sometimes be exploited by immunising the host to the fluids secreted by the feeding ectoparasite.

1.12 Inside the host

It is surprising how little has been written about the mammalian intestine from the point of view of the parasite. The most comprehensive descriptions are those of Mettrick and Podesta (Mettrick and Podesta 1974; Podesta and Mettrick 1974, 1976; Podesta 1977) of the pathophysiology of *Hymenolepis diminuta* infections in the rat. They found that the fluid content of the lumen of the small intestine was well mixed and had an oxygen tension that was similar to venous blood. This part of the intestine is thus aerobic, for such oxygen tensions are considerably greater than the threshold tension for maximal loading of respiratory cytochromes. Further, the so-called unstirred layers close to the brush border of the mucosa did not present a significant barrier to the diffusion of oxygen from the intestinal tissues into the lumen of the intestine. The distal ileum and the colon may become anoxic, but infection with *H. diminuta* appears to modify the intestinal environment in favour of the parasites. Absorption of nutrients from the lumen by the host is decreased, presumably increasing the nutrients available to the tapeworm and, at the same time, ensuring that the oxygen tension is maintained, even in the distal ileum and the colon. This is somewhat of a paradox as *H. diminuta* appears to function as an anaerobe even though it has access to oxygen in its normal environment. Infection also lowers the pH of the intestine, while the partial pressure of carbon dioxide doubles. The latter is not surprising because the intestinal buffer is bicarbonate, secreted into the intestine in large quantities in bile and pancreatic fluid. In the parasitised intestine there is thus sufficient carbon dioxide to satisfy the metabolic requirements of the parasite during anaerobic respiration. Carbon dioxide is one of the loud, unambiguous 'shouts' emanating from the host that calls forth an appropriate developmental response from the parasite (see Section 1.2). Bile may be another. Smyth (1962) observed that herbivore biles lysed *Echinococcus granulosus* protoscoleces but carnivore biles did not. He felt that it partly explained why the tapeworm was able to establish in the carnivore gut, and went on to suggest that the individual nature of bile from different types of animal provided specific signals that entrained parasite development. However, as Howell (1986) points out, cat bile does not lyse protoscoleces but the parasite fails to establish. There are obviously other determinants.

Further, there is now some evidence that parasites may modify the regulatory environment. Ovington, Barcarese-Hamilton and Bloom (1985) have shown that, in rats infected with *Nippostrongylus*, secretion of several gastrointestinal hormones markedly increased. One of them causes hyperplasia of the villi. It is interesting to speculate that this is yet another case of a parasite 'improving' its environment.

There are a number of other sites about which too little is known to enable any useful generalisations. The peritoneal cavity as an environment has been reviewed in detail by Howell (1976). Peritoneal fluid compares favourably with plasma as a source of nutrients, although it contains significantly less dissolved total solids – the difference is largely due to the greater amount of protein in plasma – and about half the oxygen (Table 1.1).

The volume of fluid in the peritoneal cavity depends on the size of the animal. Most of the space is occupied by the viscera, but fluid volumes may range from 15 to 45 ml in cattle and 0.33 to 5.58 ml in rabbits, for example. Peritoneal fluid contains large numbers of cells, including mesothelial cells from the lining of the cavity, lymphocytes, polymorphs, monocytes and macrophages. Thus, for a parasite it has disadvantages and advantages. Its main disadvantage is that the cells of the immune system are abundant. Its main advantages are that a parasite in the peritoneal cavity is in close proximity to a number of other organ systems and is not in danger of being accidentally dislodged, as it does not communicate directly with the exterior.

Table 1.1: Some Physicochemical Properties of Plasma and Peritoneal Fluid in Humans

Component	Plasma	Peritoneal Fluid
pH	7.39	7.4
solids %	8.6	2.5
proteins %	7.2	2.1
cations mEq/1	149.6	146.6
chloride mEq/1	102.4	109.0
inorganic phosphate %	4.0	4.0
sugars	81.0	14.0
pCO_2	44.9	42.4
pO_2	39-93	28-40

(Data from Howell 1976)

The vertebrate brain and the eye are frequently invaded by parasites. These sites are considered to be immunologically privileged (disadvantaged would be a better description) organs within which the immune system is not very effective, as protection depends on the integrity of the blood-brain barrier. It is a favoured haunt of larval cestodes, encysted nematodes, and errant Protozoa. Skeletal muscle also provides an environment for parasites. Unfortunately, it is not known exactly what components of the muscle system are important for the parasites, such as *Trichinella*, that inhabit it. As Despommier (1976) remarks "the physiological environment (of the muscle cell) is the sum of the physical and chemical interactions resulting in a contraction-rest cycle". *Trichinella* actually occupies the intracellular environment, a habit that it has in common with many protozoan parasites that live within a variety of cells. The best known are the leishmaniae that inhabit macrophages and the malaria parasites that live in erythrocytes. There are specific recognition molecules on surfaces of the target cells and the parasites. Unfortunately, the precise nature of these molecules and the interactions between the membranes of host cell and parasite are not well described, but the carbohydrate residues of membrane glycoproteins and glycolipids are of considerable importance.

Blood is a very special environment for parasites. It is special because it is the primary medium of the immune response and is therefore hostile. The parasite must face the problem of getting access to the blood vascular system, which is closed off from the external environment. Once access is gained, there is the additional problem of maintaining station in a continuously flowing medium. It is not important in the case of microscopic blood parasites such as the Protozoa and the microfilaria, but assumes greater significance for macroparasites like the schistosomes and the adult filaria. But it offers some very important advantages, as well. It provides an extremely stable environment – for example, in humans, it is highly oxygenated, there is ample carbon dioxide, and blood pH is regulated 'to the second decimal place'. It provides a guaranteed, constant supply of glucose and amino acids, and other growth factors, that can be absorbed directly with little metabolic alteration by the parasite.

The critical reader will have noticed that there has been little in this chapter about the environments in invertebrates that are occupied by parasites. This is because remarkably little is known about those aspects of the physiology of invertebrates that may affect their parasites. Insects and molluscs have received the most attention. In insects, the haemocoel or salivary glands are common sites for parasites, where they are bathed and supplied by the host's haemolymph. This is a nutrient medium comparable with the blood of vertebrates, but with at least one major difference, in that free glucose is not normally present. Invading parasites must therefore be able to catabolise trehalose or amino acids (e.g. proline, glutamate) for energy metabolism. This is evident, for example, in the metabolic switch that occurs during the life cycle of *Trypanosoma brucei*, when, in the bloodstream short-stumpy and insect procyclic forms, the mitochondrion develops the capacity to oxidise proline. Infection of the beetle *Tenebrio molitor* with cysticercoids of *Hymenolepis diminuta* results in the appearance of detectable glucose in the host's haemolymph; the amino acid and protein content of the haemolymph also changes (Hurd and Arme 1984). Some of these effects may be due to an interaction of the parasite with the host's endocrine system. Changes in the amino acid content have also been observed in tissues of the snail *Biomphalaria glabrata* infected with *Schistosoma mansoni* (Schnell, Becker and Winkler 1985).

The immune responses of invertebrates to parasites have been poorly studied; the major response that has been identified is that of cellular infiltration and encapsulation of the invader. It is unlikely that anything as elaborate as the responses of birds and mammals occur in invertebrates. The exquisite precision of the antibody-antigen reaction, the enhancement of the immune response by the complement system and the complicated interactions of the various classes of cells involved in immunity appear to be associated with homeothermy and, possibly, longevity of birds and mammals. Many invertebrates seem to conform to the 'throw-away' principle – protection of the individual is very much subordinated to short life cycles and the production of large numbers of individuals, of which only a few successfully reproduce. It is clearly a very effective alternative strategy to the one adopted by the larger vertebrates.

2
Energy Metabolism

2.1 Introduction

An imperative for the survival of any organism is to maintain a supply of energy for the synthesis of macromolecules and for growth, differentiation and reproduction. Parasites are, of course, no exception, and it is likely that the requirement for rapid growth and establishment in a host makes an unusually great demand on energy generating mechanisms. So, too, does the need to produce vast numbers of offspring. Protection against the immune system of the host is another costly energy drain.

Energy metabolism, then, can be defined as the sum of all those metabolic processes that lead to the net production of energy that is available to the organism. The energy is most usually made available as adenosine triphosphate (ATP). Although there is a small number of alternative energy rich compounds, such as the other nucleoside triphosphates and the acyl coenzymes A, none is as ubiquitous as ATP and, in parasites, none is as well studied.

2.1.1 *The generation of reducing equivalents*

The following brief summary of respiration in aerobic organisms is provided as a quick reference and means of comparison with the parasites that are discussed in much greater detail below. For a more complete explanation of standard biochemical processes, the reader is referred to any good biochemical text.

In wholly aerobic systems, energy is made available to the cell in a series of four stages, summarised in Figure 2.1.

In Phase 1, food (that is, potential respiratory substrates) is taken in by organisms and its constituent macromolecules are broken up, by hydrolytic enzymes, into their low molecular weight components, such as glucose or amino acids. Intracellular macromolecules used as an energy store, such as glycogen, have to be reconverted to the building blocks (glucose-6-phosphate) from which they were synthesised. Phase 1 provides no energy that can be utilised by the organism in synthetic or other processes. In Phase 2, glucose and other sugars are oxidised by the glycolytic pathway to the level of acetyl coenzyme A (Figure 2.2). A small amount of energy, in the form of ATP, becomes available during glycolysis, the enzymes of which are located in the cytosol of the cell. In aerobic organisms the

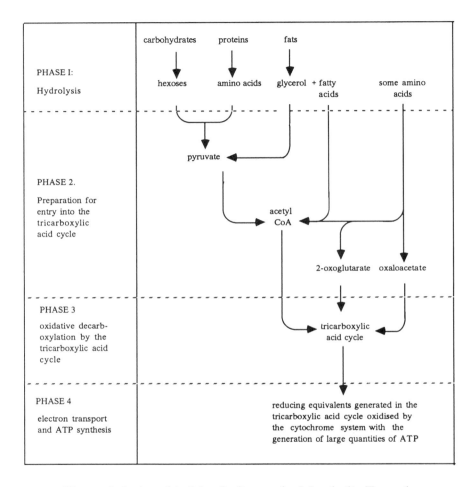

Figure 2.1: Aerobic Metabolism - the Metabolic Funnel

primary function of Phase 2 is to provide the fuel for Phase 3. Phase 3 is the tricarboxylic acid (TCA) cycle (Figure 2.3), by which the acetyl moiety of acetyl coenzyme A is oxidised to carbon dioxide and water. In the process 'reducing equivalents' are generated, in the form of reduced nicotinamide adenine dinucleotide (NADH). This process takes place in the matrix of the mitochondrion. Once again, only a small amount of energy is liberated at this stage. The reducing equivalents are transferred to the inner membrane of the mitochondrion in Phase 4. At the inner membrane, the electron transport chain, a cytochrome system, brings about their oxidation (Figure 2.4). In this process oxygen is reduced to water and ATP is synthesised using the energy of the electrochemical gradient.

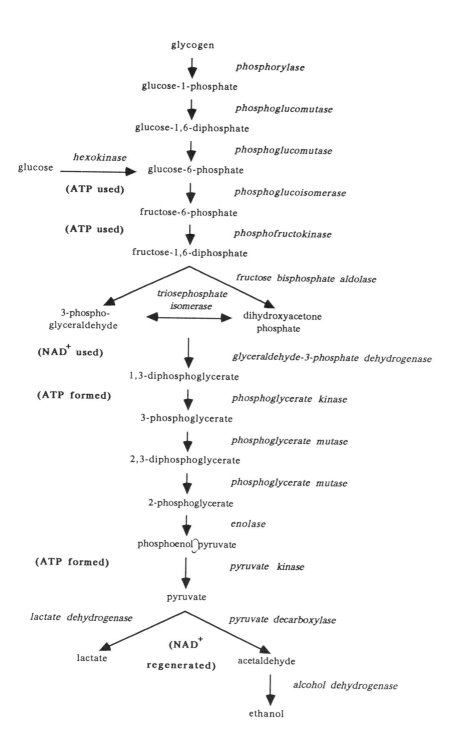

Figure 2.2: Glycolysis

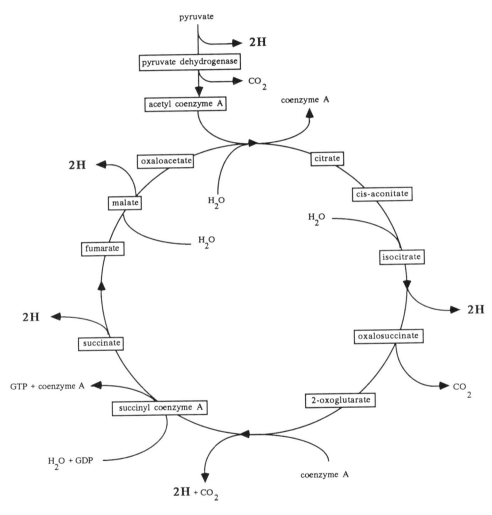

Figure 2.3: The Complete Tricarboxylic Acid Cycle

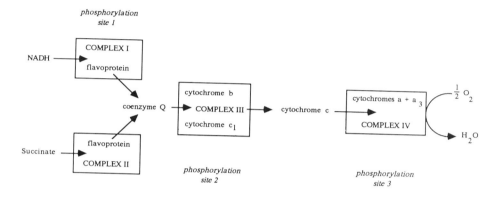

One molecule of ATP is synthesised during the oxidation of one reducing equivalent by each of the complexes I, III and IV.

NADH; reduced nicotinamide adenine dinucleotide.

Figure 2.4: The Aerobic Electron Transfer System

When a molecule of glucose is completely oxidised to carbon dioxide and water, the ATP yield is 36 molecules. This typically occurs in aerobic organisms. Parasites may be aerobic or anaerobic, and it is characteristic of them that a proportion of their respiratory substrate is not fully oxidised but is converted to fermentation products – lower fatty acids like acetic, propionic and butyric acids, or alcohols. There is, of course, an apparent cost to the parasite as the potential energy associated with those end products is lost by excretion. However, these fermentations yield more energy than glycolysis alone (Table 2.1).

Table 2.1: Some Examples of ATP Yield from Glucose in Different Types of Respiration (at Redox Balance)

End product	$\dfrac{\text{mol ATP/mol used}}{\text{mol glucose formed}}$	Organism
$CO_2 + H_2O$	36	almost all aerobes
lactic acid	2	very common
ethanol	2	many Protozoa, some helminths
alanine	2	many invertebrates
acetate	2	many Protozoa
acetate + succinate	3.7	some Protozoa, common in helminths
acetate + propionate	5.4	many helminths
acetate + propionate and other fatty acids	5	some nematodes

An organism is a steady-state system, in which energy input, in the form of respiratory substrates such as carbohydrate, must be balanced by growth, storage and energy output and loss. This requires matching catabolic processes (those which cause the breakdown of molecules) with anabolic ones (which build up molecules). Catabolic and anabolic processes are balanced through the agency of their essential cofactors and end products. For example, a high value for the ratio of the concentration of ATP to that of its hydrolysis product, adenosine diphosphate (ADP) determines that anabolic processes will be favoured over catabolic ones. When the concentration of ATP falls, due to its use in synthetic processes, anabolism is decreased in favour of catabolism. In fact, the pendulum does not swing as wildly as this would suggest because, in a healthy organism, metabolism is regulated within fairly narrow tolerances. Atkinson (1971) made use of this observation in developing the concept of the 'adenylate energy charge'. This is a measure of the energy resident in the adenylate system and is given by the following expression:

$$\frac{[ATP] + 1/2[ADP]}{[ATP] + [ADP] + [AMP]}$$

where the square brackets indicate the concentration of the adenylate in question, usually measured enzymatically in deep frozen samples of tissue that have been 'snap frozen' and extracted with perchloric acid..

The value of the adenylate energy charge is about 0.8-0.9 for many mammalian tissues; parasitic helminths give values between 0.6 and 0.9. It is a useful measure, and has frequently been employed in assessing the health of parasites in culture or in determining anthelmintic stress on a parasite. If the energy charge falls below 0.6, the animal can be considered to be moribund.

Another important determinant of metabolism is redox state which, for practical purposes, can be defined as the ratio of the intracellular concentration of NAD^+ to that of NADH. A high concentration of NAD^+ favours catabolic processes, which cannot proceed if all the NAD^+ is reduced. Barrett (1981) noted that, for *Ascaris* muscle cytoplasm, the value lies between 725 and 2214 to 1. The muscle is obviously maintained in a high state of oxidation, and this is achieved by coupling NADH oxidation to malic dehydrogenase. In the mitochondrion, however, the ratio is 0.1:1 and is probably a consequence of a mitochondrial electron transfer system based on fumarate reductase (see Section 2.3).

The two examples of regulation given above enable metabolic pathways to adjust to the overall state – the *milieu* – of the cell. But metabolic pathways are also capable of internal regulation. Product inhibition – the accumulation of an end product of metabolism preventing its further metabolism – is commonly encountered but more complex feed-back and feed-forward mechanisms are also common. These involve changing the activities of key, regulatory enzymes in the metabolic pathway.

Regulatory enzymes can easily be identified experimentally, because they have low activity, are rate-limiting in the pathway and usually catalyse reactions that are far out of equilibrium and hence are irreversible under physiological conditions. Changes in the activity of such an enzyme are therefore translated into changes in the overall activity of the pathway. An excellent example is the feed-forward activation of pyruvate kinase by fructose-1,6-bisphosphate (FBP). FBP is an early intermediate in the catabolism of glucose. If a lot of glucose becomes available to the cell the intracellular concentration of FBP rises. FBP interacts with pyruvate kinase, increasing its activity (allosteric activation) and permitting a greater carbon flow through the metabolic pathway (Figure 2.5). This is, of course, not the whole story – pyruvate kinase, like all the kinases, is also directly modulated by the concentrations of the adenine nucleotides and therefore responds to the collective effects of all those enzymes that compete for these compounds as substrates. In particular, in parasitic helminths, phosphoenolpyruvate carboxykinase is an important enzyme competing for the same substrates and cofactors (see Section 2.3).

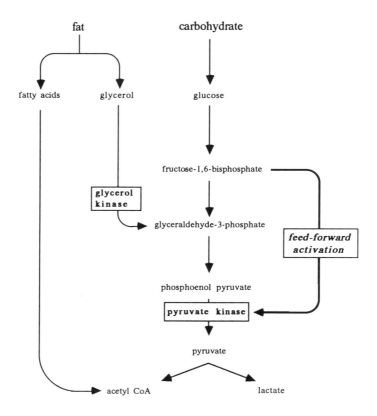

Figure 2.5: The Feed-forward Activation of Pyruvate Kinase by
Fructose-1,6-bisphosphate during Catabolism

Parasites are no different from other organisms in the way that their metabolic
pathways are regulated. They respond to adenylate energy charge, to the redox
state, their enzymes show allosteric activation similar to the FBP-pyruvate kinase
interaction just described and, generally, the same enzymes are regulatory. The
metabolic pathways are subject to feed-forward and feed-back control and to product
inhibition. In other words, they are not qualitatively different. But they are
quantitatively different. In different organisms different properties of the some
enzymes are enhanced or suppressed for the fine-tuning of metabolic pathways that
operate in different *milieus*

So far in this chapter we have concentrated almost entirely on the path of
carbon during energy metabolism but we have said little about the mechanisms by
which these reactions bring about energy conservation in a form that can be utilised

by parasites. The mechanisms of energy transduction in parasites have received much attention in the last thirty years and have been reviewed many times. For helminths, the most recent and authoritative review is that of Köhler (1985).

In general, the mechanisms by which ATP can be synthesised fall into two categories. They are substrate-linked phosphorylation and electron-transport-mediated phosphorylation.

2.1.2 Substrate-linked phosphorylation

This is a direct process, catalysed by a single enzyme or an enzyme complex, by which metabolism of a substrate molecule is coupled to the formation of a phosphate compound such as ATP. There is no evidence that the processes of substrate-linked phosphorylation in parasites are different from those of any other organism.

One class of substrate-linked phosphorylation has the general form:

$$\textit{kinase}$$
$$x\text{-phosphate} \longrightarrow x$$
$$\text{ADP} \qquad \text{ATP}$$

Examples of such reactions are afforded by:

1. Phosphoglycerate kinase
   (1,3-diphosphoglycerate + ADP $\longrightarrow$ 3-phosphoglycerate + ATP)

2. Pyruvate kinase
   (phosphoenolpyruvate + ADP $\longrightarrow$ pyruvate + ATP)

3. Phosphoenolpyruvate carboxykinase
   (phosphoenolpyruvate + CO_2 + GDP $\longrightarrow$ oxaloacetate + GTP)

GTP may then be converted to ATP in a subsequent reaction catalysed by nucleotide diphosphokinase:

$$\text{GTP + ADP} \longrightarrow \text{GDP + ATP}$$

All three reactions have been unequivocally demonstrated in parasites.

A second class of substrate-linked energy conservation involves acyl CoA transfer reactions. A good example is the formation of acetate in the malate dismutation (see Section 2.3.1). It is still not clearly understood how this occurs, but the most favoured explanation is that acetyl CoA is first produced from pyruvate by the action of the pyruvate dehydrogenase complex and the CoA is then transferred to succinate, which is a product of the other arm of the malate dismutation:

$$\text{acetyl CoA} + \text{succinate} \longrightarrow \text{acetate} + \text{succinyl CoA}$$
(acetyl CoA transferase)

succinyl CoA is then involved in a substrate-linked phosphorylation, thus:

$$\text{succinyl CoA} + \text{GDP} + P_i \longrightarrow \text{succinate} + \text{CoA} + \text{GTP}$$
(succinyl CoA synthetase)

And finally, it appears that both *Entamoeba histolytica* and *Giardia intestinalis* are able to couple ATP formation directly to the cleavage of the thioester bond:

$$\text{acetyl CoA} + \text{ADP} + P_i \longrightarrow \text{acetate} + \text{CoA} + \text{ATP}$$

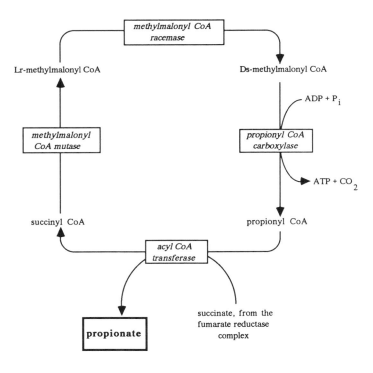

Figure 2.6: The Succinate Decarboxylase System

Köhler (1985) points out that acyl CoA transferase reactions and acyl CoA-dependent ATP conservation play an important role in energy metabolism in parasites. Propionate formation is a case in point and occurs in *Fasciola* and *Ascaris* as illustrated in Figure 2.6.

The formation of propionate is another good example of the reversal, in parasites, of a process found in free-living organisms – that is, the utilisation of propionate as a substrate for gluconeogenesis in ruminants.

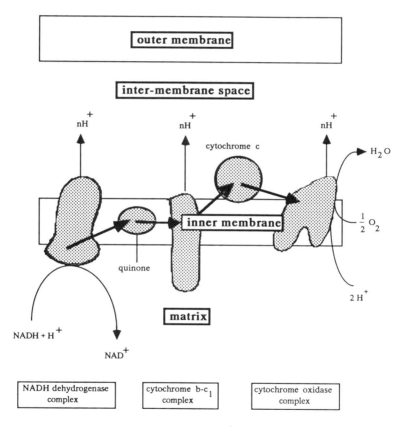

As two electrons are transferred from NADH + H^+ to oxygen, the orientation of each complex in the inner mitochondrial membrane is such as to transfer protons from the matrix to the inter membrane space.

Figure 2.7: Electron Flow through the Respiratory Complexes in Mammalian Mitochondria

2.1.3 *Electron transport mediated phosphorylation*

The classical, mammalian electron transport system was briefly summarised at the beginning of this chapter (Figure 2.4). It involves four complexes, three of which catalyse proton translocation during the oxidation of one reducing equivalent of NADH. The complexes are numbered as follows: I, the flavin-based NADH dehydrogenase; II, the flavin based succinate dehydrogenase; III, the cytochrome b + c_1 complex; and IV, the cytochrome a + a_3 complex. Electron transfer between complex I and III or II and III is mediated by a carrier called ubiquinone. The components of this system are arranged in the inner mitochondrial membrane. NADH dehydrogenase and succinate dehydrogenase are exposed on the inner surface of the membrane, providing access to NADH and succinate respectively, whereas cytochrome c is an extrinsic protein associated with the outer surface of complex IV. Electron transport and the reduction of oxygen in complex IV take place towards the cytosolic side of the complex (Figure 2.7).

   It is considered that the arrangement of the components of the electron transport system permits it to act vectorially. That is, as each of the electron carriers is successively reduced and oxidised, protons are selectively transferred to the outer side of the membrane, i.e. into the intermembrane space, without accompanying anions. Proton transfer lowers the pH of the fluid in the intermembrane space, setting up a pH gradient and a charge difference across the inner mitochondrial membrane. The 'proton motive force' thus set up is released by another protein set in the inner mitochondrial membrane with its active centre on the inner surface. This protein is the F_1F_O ATP synthetase, also known as complex V. It provides a channel for protons to flow back across the membrane to the inner surface. The proton flow derives its potential energy from the electrochemical and the pH gradients across the membrane. The energy associated with the proton flow drives ATP synthesis. Köhler (1985) depicts this mechanism for generating a proton motive force in helminths. In general terms it applies to anaerobes as well as aerobes (Figure 2.8).

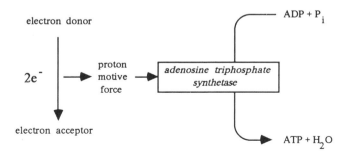

Figure 2.8: Mitochondrial ATP Synthesis Driven by a Proton Motive Force

The interested reader is directed to any modern biochemical text for a complete treatment of this very complex process. We discuss it here to permit comparisons between the aerobic and anaerobic systems of parasites.

2.2 Energy metabolism in Protozoa

Protozoan parasites have evolved from a wide variety of taxonomic groups and most have several stages in their life cycles that involve a metabolic 'switch' on changing from one stage to the next. The result is that energy metabolism among the protozoan groups and stages is quite diverse and we are unable here to do justice to the many subtle variations that exist.

Glucose is the major substrate for energy metabolism in those parasitic groups and stages that do not utilise the tricarboxylic acid cycle to any great extent, such as the malarial parasites of mammals, bloodstream salivarian trypanosomes, trichomonads, *Giardia* and *Entamoeba*. Protozoan groups and stages with an active tricarboxylic acid cycle (*Trypanosoma cruzi*, *Leishmania* spp, procyclic salivarian trypanosomes) catabolise fatty acids and amino acids, especially proline, in stages that infect insects. However, respiration is seldom fully oxidative in any of these groups.

Glycolysis to lactate is the major energy-producing pathway in intraerythrocytic stages of the *Plasmodium* and *Babesia* species that infect mammals (see Sherman 1984), though there is *in vitro* evidence for a minor mitochondrial contribution in *P. falciparum* (Ginsburg, Divo, Geary, Boland and Jensen 1986). In contrast, those *Plasmodium* species that infect birds, or other vertebrates with nucleated red blood cells, appear to have full tricarboxylic acid cycle activity and excrete CO_2 as well as lactate. Fully-developed tricarboxylic acid cycle activity is also found in intracellular parasites such as *T. cruzi* and amastigote stages of *Leishmania* spp. *T. cruzi* nonetheless maintains a fumarate reductase (Boveris, Hertig and Turrens 1986). This is also the case for *Leishmania* species – for example, *L. braziliensis* produces succinate, acetate, pyruvate, D-lactate and alanine under aerobic conditions and glycerol anaerobically (Darling, Davis, London and Blum 1987). Interestingly, this is the only protozoan parasite reported to date to produce D-lactate. *Entamoeba*, *Giardia* and the trichomonads do not possess mitochondria and rely on unique special pathways (discussed below) for ATP generation.

Special regulatory processes and structural variations are found in some of the bioenergetic pathways of Protozoa and these may change with different stages of the life cycle. For example, *L. donovani* amastigotes, which inhabit the phagolysosomes of host macrophages, carry out most of their metabolic activities best at acid pH (4-5.5), whereas the same activities in the promastigotes, which are not intracellular, function optimally at neutral pH (Mukkada, Meade, Glaser and Bonventre 1985). The distribution and activities of glycosomal and soluble enzymes in *T. brucei* and *Leishmania* spp. vary with the stage of the life cycle

(Opperdoes and Cottem 1982; Coombs, Craft and Hart 1982; Blum 1987). There is evidence that the regulatory properties of phosphofructokinase, an enzyme important in the regulation of glycolytic flux in many organisms, are different in trypanosomes from other organisms, and that the properties also vary between stages of the life cycle (Cronin and Tipton 1985; Taylor and Gutteridge 1986). Phosphoenolpyruvate carboxykinase in most parasitic Protozoa is different, too, in that like other groups of microorganisms and plants, it requires ADP instead of GDP or IDP.

2.2.1 *Special pathways in Protozoa*

A variety of apparently unique pathways for energy metabolism is found in several protozoan groups and these are often associated with unusual cellular organelles, such as the glycosomes in kinetoplastids and the hydrogenosomes in trichomonads.

Aerobic glycolysis in bloodstream trypanosomes
Trypanosoma brucei is the most-studied model of metabolism in bloodstream trypanosomes and provides what appears to be the most clear-cut example of metabolic adaptation in parasites that we can identify. However, as Coombs (1986) points out, the metabolism of *T. brucei* may not be typical of all bloodstream trypanosomes.

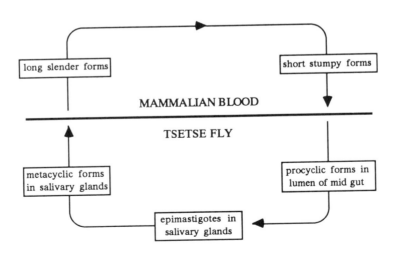

Figure 2.9: Life Cycle of *Trypanosoma brucei*

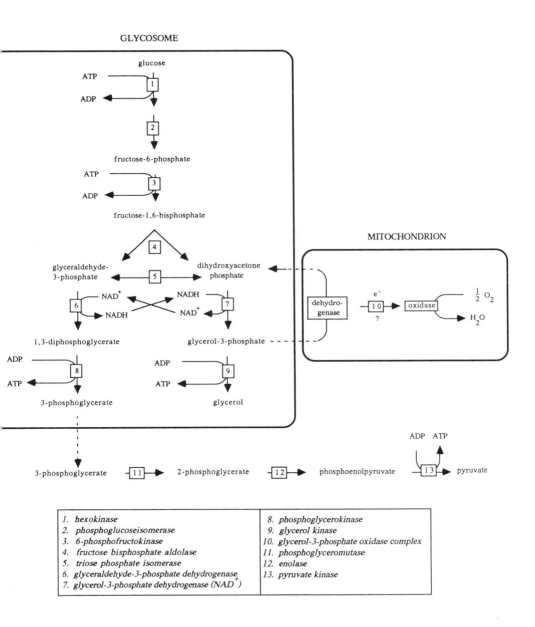

Figure 2.10: Aerobic Glycolysis in *Trypanosoma brucei* Long-Slender Bloodstream Forms (after Fairlamb 1982)

The pathways of energy metabolism in *T. brucei* vary with the stage of the life cycle, which is outlined in Figure 2.9 (see Fairlamb 1982). The blood trypomastigotes, which are the long slender forms that first appear in the bloodstream after infection, catabolise glucose by a pathway that has been termed 'aerobic glycolysis'. This pathway, depicted in Figure 2.10, is unique to the bloodstream stages of African trypanosomes. It involves interaction between two organelles and the cytosol. The glycosome contains all the enzymes of glycolysis to the level of phosphoglycerate kinase, the cytosol contains the terminal enzymes, phosphoglycerate mutase and enolase, and the mitochondria contain the glycerophosphate oxidase complex. There is no lactate dehydrogenase activity. At this stage of the life cycle the single mitochondrion, which is a simple tube, possesses no tricarboxylic acid cycle or classical electron transport activity (Brown, Evans and Vickerman 1973).

Glucose is taken up rapidly from the bloodstream and metabolised immediately to the end product pyruvate. Interestingly, these organisms show no anomeric specificity for α- or β-glucose (Mackenzie, Hall, Flynn and Scott 1983). Bloodstream trypanosomes do not store carbohydrates and, when oxygen is not limiting, glucose transport is the rate-limiting factor in the pathway. Redox balance is maintained within the glycosome by the glyceraldehyde-3-phosphate dehydrogenase / glycerol-3-phosphate dehydrogenase enzyme pair. Glycerol-3-phosphate and dihydroxyacetone phosphate shuttle to and from the mitochondrion where glycerol-3-phosphate is oxidised by the glycerophosphate oxidase complex which reduces oxygen to water. Pyruvate is not further metabolised and there appears to be no phosphorylation associated with the mitochondrial electron transfer steps. Thus, the potential yield of ATP from this pathway is only 2 moles/mole glucose. *In vitro*, when long slender forms are incubated under anoxic conditions, glycerol is produced in a 1:1 ratio with pyruvate, with a reduced potential yield of ATP (1 mole/mole glucose) but it appears unlikely that this pathway operates to any great extent at the oxygen concentrations prevailing *in vivo* (Eisenthal and Panes 1985).

It is interesting to consider the origins or function of aerobic glycolysis, because in energetic terms it provides no apparent advantage over fermentation to lactate and, under anoxic conditions, in fact yields less ATP than lactate production. It would be interesting to know what effect pyruvate excretion by large numbers of bloodstream trypanosomes has on the host. Clearly, bloodstream trypanosomes, living in an environment with a relatively high and constant concentration of glucose and having an extraordinarily high rate of utilisation of glucose, do not need to maximise their yield of ATP and hence can afford to be profligate in the utilisation of their energy sources. But glycolysis to lactate is an ancient pathway, utilised by many other blood-dwelling parasites, and it is difficult to see a selective evolutionary advantage in favour of the elaboration of aerobic glycolysis, unless it is in fact an evolutionary remnant of an ancestral pathway that originally did confer energetic advantages.

Glycosomes are microbodies surrounded by a single membrane that are found in all the kinetoplastids and are unique to this group. They are probably related to peroxisomes (Opperdoes 1987). They contain high concentrations of nine of the glycolytic enzymes, plus several other enzymes. Glycosomes are found in all kinetoplastids that have been examined, at all stages of the life cycle, including those with high tricarboxylic acid cycle activity. The origins of these organelles are lost in evolutionary history, but it appears that they can sustain a very high rate of glycolysis because of the high concentration of glycolytic enzymes and the permeability barrier to large molecules presented by the limiting membrane (Patthey and Deshusses 1987). There is no evidence for a postulated substrate channelling in glycosomes by multienzyme complexes (Aman and Wang 1986).

The glycerophosphate oxidase complex, which is unique to African trypanosomes, is located in the mitochondrial membrane. It consists of two major components, an FAD-linked dehydrogenase and a terminal oxidase that may contain copper. There is evidence also for an intermediary ubiquinone-like molecule (Turrens, Bickar and Lehninger 1986). Oxygen uptake is insensitive to cyanide but sensitive to the metal chelator salicylhydroxamic acid.

The role of the mitochondrion in energy metabolism changes at different stages of the life cycle. In long slender bloodstream forms, the simple tubular mitochondrion contains no cytochromes or haem and the glycerophosphate oxidase complex is responsible for respiration. During the course of infection in the mammalian host, however, different morphological bloodstream forms appear, 'intermediate' forms, which are capable of division, and short-stumpy forms, which do not divide. Recent *in vitro* evidence suggests that the intermediate forms, but not the short-stumpy forms, are infective to the tsetse fly (Giffin, McCann, Bitonti and Bacchi 1986). The mitochondrion in both these forms is more elaborate, with some branches and cristae. They are able to metabolise pyruvate to acetate or succinate and contain fumarate reductase and all the tricarboxylic acid cycle enzymes, although the cycle does not function. During this transformation process, enzyme activities in the glycosome change: glycolytic enzymes decline while phosphoenolpyruvate carboxykinase and malate dehydrogenase activities appear (Hart, Misset, Edwards and Opperdoes 1984). At this stage glycosomal NADH is reoxidised via malate dehydrogenase.

The intermediate forms are preadapted for survival in the insect gut. After ingestion they gradually transform to procyclic trypomastigotes and migrate to the salivary glands. At this stage they have a highly branched mitochondrion, with large numbers of cristae, and a fully functional tricarboxylic acid cycle. They are able to oxidise proline and other amino acids in the insect's haemolymph and oxygen uptake is partially cyanide-sensitive. But their metabolism is not fully oxidative, as they excrete succinate, pyruvate and acetate as well as CO_2 as end products of glucose catabolism. The metabolism of *T. cruzi* is similar to that of the procyclic forms of *T. brucei*, with little variation during the life cycle.

In the metabolic changes that occur in the parasites during the course of the life cycle, we have an example of what appears to be close adaptation to environmental

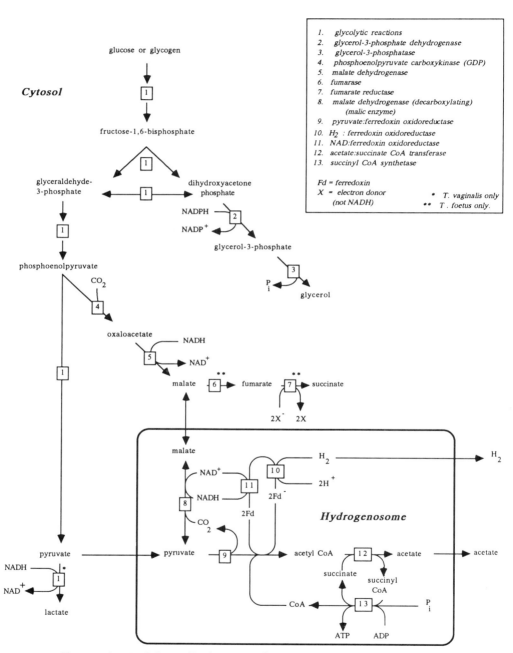

Figure 2.11: Major Pathways of Anaerobic Glucose Catabolism in Trichomonads (modified from Steinbuchel and Muller 1986a)

conditions at each stage, in that the glucose-based pathways of the bloodstream forms give way to the amino acid-based oxidative pathways of the insect forms. But we are still far from understanding the evolutionary and adaptational bases of aerobic glycolysis in the bloodstream forms and of the cyanide-insensitive respiration of the insect forms.

Hydrogenosomal pathways in Trichomonads

Trichomonads are aerotolerant anaerobes. They do not possess mitochondria and utilise carbohydrates (stored as glycogen) as their major substrate. They possess unique intracellular organelles, the hydrogenosomes, which contain enzymes catalysing the terminal sequences of carbohydrate catabolism. The hydrogenosome is not a modified mitochondrion: although it has a double membrane, it possesses no DNA, no cytochromes or DCCD-sensitive ATPase and the inner membrane does not contain cardiolipin (Müller 1980; Turner and Müller 1983; Paltauf and Meingassner 1982). Peroxisomal enzymes have not been detected in these organelles.

The major pathways of anaerobic glucose catabolism in trichomonads are summarised in Figure 2.11. The precise pathways vary slightly between species and strains. Glycolytically-formed pyruvate and possibly malate are the oxidisable substrates for hydrogenosomal metabolism. The end products formed are acetate, hydrogen, glycerol, lactate (*T. vaginalis* only) and succinate (*T. foetus* only). In *T. foetus* the extent of malate formation in the cytosol or hydrogenosome is influenced by pCO_2. A substrate-level phosphorylation is associated with acetate production, which demonstrates an energetic advantage in the development and maintenance of hydrogenosomal metabolism. The formation of glycerol is important for supplying reducing equivalents to support hydrogenosomal metabolism, especially in *T. vaginalis* (Steinbüchel and Müller 1986a).

Hydrogenosomes occupy a large volume of the cytoplasm in trichomonads and the flow of carbon through these organelles is substantial. In *T. foetus* up to 50% of glycolytically generated pyruvate is processed in the hydrogenosome (Steinbüchel and Müller 1986b).

An essential component of the hydrogenosomal pathways is the pyruvate synthase enzyme complex which oxidatively decarboxylates pyruvate to acetyl CoA. Electrons are transferred to a ferredoxin by pyruvate:ferredoxin oxidoreductase, an extrinsic hydrogenosomal membrane protein, and thence to protons via a H_2:ferredoxin oxidoreductase (hydrogenase), forming hydrogen. Electrons may also be transferred to NAD^+ via a NAD:ferredoxin oxidoreductase, generating NADH to support the malic enzyme reaction. High concentrations of CoA, NAD^+ and ferredoxin are maintained within the hydrogenosome, which presents a permeability barrier to passage of these cofactors across the membrane.

Hydrogen is not produced when oxygen is present to act as an electron acceptor. Trichomonads utilise oxygen at a high rate, increasing the production of acetate in preference to the cytosolically-formed end products. The majority of oxygen reduction takes place in the cytosol, catalysed by an active, soluble NADH oxidase which has iron-sulphur and flavin components. This oxygen-scavenging activity, which is insensitive to cyanide, may have a detoxification role in these parasites, minimising access of oxygen to the hydrogenosome and the resulting inactivation of enzymes and anaerobic electron transport processes. The activities of NADH oxidase in different trichomonad species vary, perhaps reflecting the ambient oxygen levels and hence required levels of detoxification, in their respective environments (Thong and Coombs 1987).

The properties of purified pyruvate:ferredoxin oxidoreductase have been shown to be similar to those of equivalent enzymes in certain bacteria except that the trichomonal enzyme is membrane-bound (Williams, Lowe and Leadlay 1987). The enzyme has two (4Fe-4S) clusters and contains thiamine pyrophosphate. The presence also of hydrogenase and other enzymes with bacterial properties suggests a possible evolutionary origin of this organelle from bacteria. This acquisition appears to be an evolutionary event that was confined to this group of eukaryotes.

Pathways in Entamoeba and Giardia
Many aspects of the metabolism of members of the genera *Entamoeba* and *Giardia* are unique and have some similarities with metabolism in the trichomonads. *Entamoeba* and *Giardia* do not possess mitochondria or other organelles dedicated to energy metabolism. Many characteristics of the metabolism of *Entamoeba* are allied more to prokaryotes or primitive eukaryotes than is the case with other groups of Protozoa. *Entamoeba* does not have a nucleolus, normal ribosomes, microtubules or the well-developed membrane structures typical of a eukaryotic cell, such as a true Golgi apparatus or an extensive endoplasmic reticulum (McLaughlin and Aley 1985).

Investigation of metabolism in *Entamoeba* trophozoites and, to some extent, that of *Giardia* , too, has been complicated in the past by the presence of ingested bacteria and their associated enzymes, which may, in monoxenic or polyxenic culture or under field conditions, contribute significantly to the metabolism of the parasite. For example, glutathione metabolism is absent from axenically-cultured *Entamoeba* but is present in organisms from monoxenic cultures (Fahey, Newton, Arrick, Overdank-Bogart and Aley 1984). For a long time it was considered that *Entamoeba* possessed hydrogenase activity, as hydrogen had been shown to be an excretory product, but more recently this has been attributed to the bacterial enzyme. There is no doubt that exogenous enzymes are active in *Entamoeba* in its natural state. To what extent, if any, these enzymes are integrated into the metabolism of the host amoebae is not clear. This raises interesting questions about the origins of the unique enzymes (discussed below) found in *Entamoeba* .

Entamoeba species are aerotolerant anaerobes (see Reeves 1984a; McLaughlin and Aley 1985, for reviews). They can consume oxygen when it is present, utilising either the highly active NAD(P)H oxidases present in the cytosol, or reactions associated with the pyruvate synthase complex. They require glucose (or galactose) as the principal respiratory substrate and possess specific transport processes for glucose. Indeed, the rate of glucose uptake appears to be the rate-limiting factor in its utilisation. A specific glucokinase converts glucose into glucose-6-phosphate and considerable quantities are stored as glycogen. The enzymes of glycogen synthesis and glycogenolysis lack the regulatory features found in animal cells. A glycogen cycle (see Figure 2.12) may contribute pyrophosphate which is present in high concentrations in *Entamoeba* cytosol and is important in glucose catabolism. Glucose-6-phosphate dehydrogenase activity is reported to be absent in these parasites, indicating that NADPH is not generated by this route.

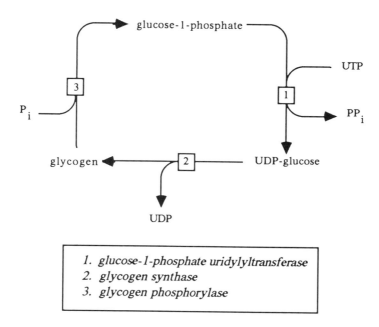

Fig 2.12: The Glycogen Cycle as a Source of Pyrophosphate in *Entamoeba*

The end products of glucose catabolism are ethanol and CO_2 under anaerobic conditions and acetate plus ethanol and CO_2 aerobically. No tricarboxylic acid cycle activity has been detected. The pathways of glucose utilisation in *Entamoeba* are shown in Figure 2.13. There is no compartmentation of glycolysis in this group. The pathways have some unusual features. The phosphorylation of fructose-6-phosphate is catalysed by a pyrophosphate-dependent phosphofructokinase, an enzyme also found in certain bacteria and plants. It has different properties from the

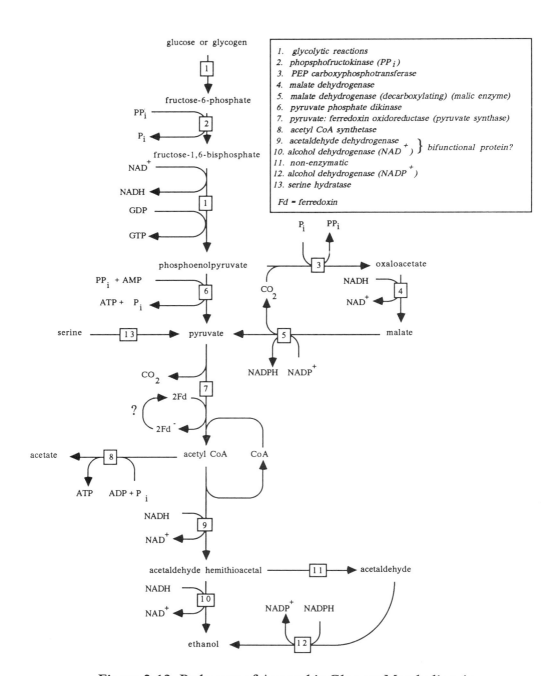

Figure 2.13: Pathways of Anaerobic Glucose Metabolism in *Entamoeba*

ATP-dependent enzyme and is not controlled by the regulatory processes that control the ATP-dependent enzyme. The activity of the *Entamoeba* enzyme is fully reversible, unlike ATP-dependent phosphofructokinase. No fructose-1,6-bisphosphatase activity has been detected.

The metabolism of phosphoenolpyruvate is unusual. Pyruvate kinase and phosphoenolpyruvate carboxykinase activities are not present. Pyruvate may be formed by either a pyrophosphate-dependent pyruvate phosphate dikinase or by a combination of phosphoenolpyruvate carboxyphosphotransferase (pyrophosphate-dependent) plus malate dehydrogenase and malic enzyme, thus recycling glycolytic NAD^+ and generating an equivalent of NADPH. Pyruvate phosphate dikinase is found in bacteria and plants but has not been reported from other Protozoa or from animals cells. Similarly, phosphoenolpyruvate carboxyphosphotransferase has been reported from propionibacteria but amongst the eukaryotes it is unique to *Entamoeba*.

Pyruvate is converted into acetyl CoA by a pyruvate synthase that donates electrons to a low redox potential iron-sulphur protein, probably ferredoxin (Figure 2.14). Aerobically, oxygen may serve as the terminal electron acceptor, though the precise electron transport pathway has not yet been completely elucidated (Weinbach 1981). Cytochromes have not been detected in *Entamoeba*. Anaerobically, the ultimate electron acceptor for this enzymatic step is not known. Hydrogenase, which serves this function in trichomonads, is not present in axenic *Entamoeba* but may be important under natural conditions. Under aerobic conditions acetate is produced in addition to ethanol and CO_2. The acetyl CoA synthetase responsible for catalysing this conversion synthesises ATP, not pyrophosphate.

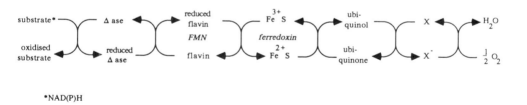

*NAD(P)H

Δ ase = dehydrogenase

Figure 2.14: Putative Aerobic Respiratory Chain in *Entamoeba* (modified from Weinbach 1981)

Acetyl CoA is reduced to acetaldehyde by acetaldehyde dehydrogenase and a second reduction of the acetaldehyde, catalysed by alcohol dehydrogenase, forms ethanol. Acetaldehyde is not released from acetaldehyde dehydrogenase but remains conjugated as a hemithioacetal, which is the true substrate for NADH-dependent alcohol dehydrogenase. A small proportion of free acetaldehyde is also produced, which is reduced by NADPH-dependent alcohol dehydrogenase.

It has been noted by Reeves (1984b) that the metabolic pathways outlined above have no regulatory points other than the initial uptake of glucose and the thermodynamic equilibria of the conversion steps. This is interesting and unusual and raises the question of evolutionary development. Does this 'unregulated' pathway represent that of a very primitive organism or is it a result of loss of regulation due to the parasitic mode of life? As Reeves has pointed out, the pathway, with its reliance on pyrophosphate and thiol compounds for energy conservation, could be a relic from a very early stage in the evolution of eukaryotes that has been conserved because of its relative isolation in a parasitic mode of life. This idea is supported by the relative ease with which *Entamoeba* appears to acquire and utilise exogenous enzymes from associated organisms.

Although there are some similarities between the metabolism of *Entamoeba* and that of *Giardia*, which are not closely related taxonomically, insufficient work has been carried out on *Giardia* for detailed comparisons to be made. *Giardia* does not possess the unique enzymes found in *Entamoeba* , although it produces similar end products and utilises a pyruvate:ferredoxin oxidoreductase in the synthesis of acetyl CoA. The apparent metabolic similarities between these groups (and also the trichomonads) may be due to their occupation of similar habitats.

Giardia parasites are aerotolerant anaerobes that do not possess mitochondria or other organelles associated with energy metabolism. Anaerobically they metabolise glucose or stored glycogen to ethanol, acetate and CO_2 (Lindmark 1980). More acetate is formed when oxygen is present. There is no tricarboxylic acid cycle activity (Weinbach, Clagett, Keister, Diamond and Kon 1980). *Giardia* consumes oxygen when it is available but, *in vitro*, grows best when oxygen concentrations are low. Oxygen uptake is insensitive to cyanide and the parasites contain no detectable cytochromes or covalently-bound flavonucleotides. Low redox potential iron-sulphur electron carriers are present. The proposed pathways for glucose degradation in *Giardia* are presented in Figure 2.15. The activities of this pathway are all found in the soluble fraction of the cells. The terminal steps in ethanol synthesis are unclear at present. An active NAD(P)H diaphorase is present in particulate fractions and H_2O rather than H_2O_2 is the product of oxygen reduction. As has been postulated for trichomonads and *Entamoeba* , this activity may be a detoxification mechanism.

2.2.2 Electron transport in parasitic Protozoa

The 'anaerobic' protozoan groups, represented by *Entamoeba* , *Giardia* and the trichomonads, do not possess mitochondria and do not appear to have haem proteins such as cytochromes or catalase. They therefore have no direct equivalent of the mammalian mitochondrial respiratory chain. The mediators of electron transport in

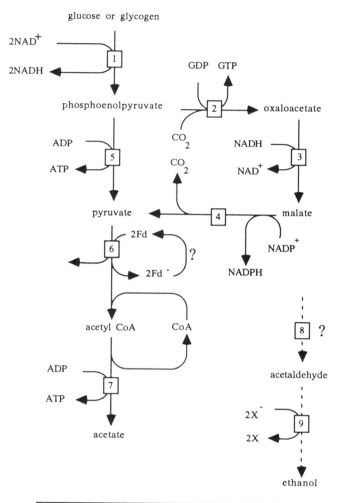

Figure 2.15: Putative Pathways of Anaerobic Glucose Catabolism in *Giardia* (modified from Lindmark 1980)

these groups are quite different from those in other parasites and are adapted to function under the relatively anaerobic conditions in which the parasites live. In particular, these include low redox potential iron-sulphur compounds with the properties of ferredoxins which are active in both anaerobic and aerobic electron transport.

Anaerobic electron transport in these groups is linked to the oxidation of pyruvate by pyruvate:ferredoxin oxidoreductase and in those species examined the initial electron acceptor is a ferredoxin. Subsequent transfer steps vary between groups: in *Giardia* and *Entamoeba* they have not been fully identified; in trichomonads electrons are transferred from ferredoxin to protons, producing hydrogen. Phosphorylation of ADP is not associated with this system in any of these groups.

These 'anaerobic' Protozoa are also capable of reducing oxygen to water, using NAD(P)H as the electron donor. In this case, the initial electron acceptors are non-covalently bound flavins, identified as flavin mononucleotide in the case of *Entamoeba* Additional, unidentified, electron carriers are also present. The inhibitors antimycin, azide and cyanide have no effect on respiration in these organisms. A tentative scheme for aerobic electron transport in *Entamoeba* is shown in Figure 2.14. It is unlikely that energy conservation is associated with this respiratory pathway. It should be noted that ubiquinone occurs in very low concentration in *Entamoeba*, so this scheme is by no means definitive. The fact that TMPD (*N,N,N',N'-* tetramethyl-*p*- phenylenediamine) is able to donate electrons at (presumably) the ultimate step is interesting in that it shows that there is a (non-haem) carrier present with redox properties similar to those of cytochrome c.

The other major groups of protozoans, the trypanosomes, leishmaniae and plasmodia all possess mitochondria containing haem compounds in the form of cytochromes. They all respire aerobically but the properties of their respiratory chains vary between groups and stages of the life cycle. All appear to possess some proportion of cyanide-insensitive terminal oxidase activity and, like the helminths, may have branched electron transport chains.

In culture forms of *T. brucei* , which are equivalent to the insect procyclic forms, about 60% of respiration is sensitive to inhibition by cyanide and antimycin A, the residue being sensitive to salicylhydroxamic acid (Bienen, Hill and Shin 1983). Bloodstream forms, on the other hand, which contain no cytochromes, are sensitive only to salicylhydroxamic acid and possess a simple respiratory chain transferring electrons from glycerol-3-phosphate dehydrogenase via a ubiquinol to the terminal oxidase. There appears to be no energy conservation associated with this pathway in the bloodstream forms.

T. cruzi epimastigotes contain three terminal oxidases: the major activity is due to cytochrome aa3, which is sensitive to low concentrations of cyanide, and to two other minor oxidases, one of which appears to be a cytochrome o (Affranchino, Schwarcz de Tarlovsky and Stoppani 1986). Respiratory control and coupling sites II and III have been demonstrated in the cyanide-sensitive pathway, but, interestingly, the pathway was not sensitive to the site I inhibitor rotenone (Affranchino, De Tarlovsky and Stoppani 1985). A cytosolic FAD-containing NADPH:cytochrome c reductase has been purified form *T. cruzi* (Kuwahara, White and Agosin 1985). This enzyme has some similarities to microsomal NADPH:cytochrome P-450 reductase and may have a detoxification function in *T. cruzi*.

Respiration of *L. mexicana* amastigotes and promastigotes is sensitive to cyanide and antimycin but insensitive to salicylhydroxamic acid, indicating the presence of a classical-type respiratory chain in this and other members of this group (Hart, Vickerman and Coombs 1981). It is not known at present whether the pathway is able to catalyse oxidative phosphorylation.

A branched electron transport chain with two terminal oxidases, a cyanide-sensitive cytochrome oxidase and a salicylhydroxamic acid-sensitive oxidase, has been proposed for *Plasmodium* species (see Sherman 1984) but the role of respiration in mammalian malaria parasites is enigmatic, since the catabolism of glucose in these organisms yields lactate as the only end product. It appears likely that oxygen utilisation is associated instead with dihydroorotate dehydrogenase activity (Gutteridge, Dave and Richards 1979). Electrons from dihydroorotate are probably transferred to ubiquinone and thence to a terminal oxidase.

2.3 Energy metabolism in helminths

All helminths so far studied utilise glucose as a respiratory substrate. Adult cestodes *in vitro* are capable of absorbing all the detectable glucose from their incubation media. This indicates that the parasites have a very active transport mechanism which binds glucose at very low concentrations and transfers it to the cytosolic compartment. Once glucose is absorbed, it is either converted into glycogen (or other storage carbohydrates), to act as an energy store – helminth parasites contain surprisingly large reserves of glycogen – or metabolised directly via the glycolytic sequence of reactions as far as phosphoenolpyruvate. This sequence is the same as that found in free-living organisms, and appears to have similar regulatory characteristics. As remarked earlier, there is no evidence that control of catabolic pathways in parasites differs fundamentally from that in free-living organisms; however, there is evidence that some components of the pathway interact in somewhat different ways.

For example, a recent study of the dog tapeworm, *Taenia serialis* (Kohlhagen and Bryant, in preparation), suggests that, like many other organisms, it is able to use glycerol for energy generation, oxidising it, after an initial phosphorylation step, via the latter half of the glycolytic pathway. Fructose-1,6-bisphosphate (FBP), the intermediate of glycolysis responsible for the feed-forward activation of pyruvate kinase, is produced at a point in the pathway before entry of glycerol (Figure 2.5). Pyruvate kinase from *T. serialis* is not activated by FBP. A possible explanation for this is that, in a glycerol-metabolising tapeworm, FBP is not a good indicator of carbon flux. Thus, cestodes that rely on glycerol as a respiratory substrate may have pyruvate kinases that do not require FBP for maximum activity. The enzyme from *Moniezia expansa*, however, *is* activated by FBP. *M. expansa* is an intestinal cestode in sheep and encounters very little glycerol in its normal environment. We are therefore tempted to speculate that pyruvate kinases in tapeworms from ruminants will be regulated by FBP, while those from carnivores that have a large component of fat in their diets, will not be.

2.3.1 *Special pathways in helminths*

Until recently, it has been considered that the respiratory metabolism of parasitic helminths conformed to three main patterns, which are illustrated in Figures 2.16, 2.18 and 2.20. These patterns are quite distinct, although as Barrett (1989) has pointed out, the third is really an extension of the second, which is the fundamental anaerobic pathway that is encountered in the majority of anaerobic or partially anaerobic metazoans.

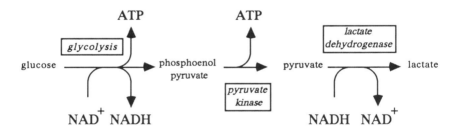

Figure 2.16: Homolactate Fermentation

Homolactate fermentation
Homolactate fermentation (Figure 2.16) is supposedly well exemplified by the filarial worms and the schistosomes. Strictly speaking, in homolactate fermentation glucose is converted exclusively into lactic acid by glycolysis. This process is restricted to the cytosol. The end product, lactic acid, is excreted and energy generation is thus wholly independent of oxygen. Energy, as ATP, is trapped in the reactions catalysed by phosphoglycerate kinase and pyruvate kinase. The pathway remains in redox balance as the reducing equivalents (NADH) generated in the early stages of the pathway are reoxidised by the terminal step, catalysed by lactate dehydrogenase.

There is no evidence that any parasitic helminth exhibits homolactate fermentation in the strict sense of the word (Bryant, 1988). It implies that the parasite is solely dependent on the lactate-producing pathway and that no other energy-generating pathways exist. There are now several, very detailed, studies that conclude that there is considerable evidence for aerobic respiration in schistosomes and the filarial worms. Huang (1980) showed experimentally and by reexamining the work of others that *Schistosoma japonicum* has a substantial capacity for aerobic respiration. Van Oordt, Van den Heuvel, Tielens and Van den Bergh (1985) have also calculated, for *Schistosoma mansoni*, that even the low levels of tricarboxylic acid cycle activity that are indicated by the generation of small amounts of carbon dioxide from glucose under aerobic conditions could make a substantial contribution to energy metabolism. This is because aerobic catabolism of glucose generates eighteen times as much ATP, per mole of glucose oxidised, as glycolysis.

Similar amounts of tricarboxylic acid cycle activity, also detected by carbon dioxide generation during aerobic incubation *in vitro*, have long been observed for filarial worms. *Brugia pahangi* and *Dipetalonema viteae* possess mitochondria whose inner membranes are thrown into folds or cristae. Such folding usually indicates that the mitochondria are capable of aerobic respiration. Further, the adults of both parasites show significant oxygen consumption. The oxygen uptake (as well as the motility of the worms) is completely inhibited by cyanide, while rotenone and antimycin A (inhibitors of complexes I and III in the mammalian electron transport system) suppress it by about 80% (Barrett 1983). Compounds that 'uncouple' oxidative phosphorylation, by destroying the link between ATP synthesis and electron transfer, usually cause oxygen uptake to accelerate. These uncouplers stimulate endogenous oxygen consumption in *B. pahangi* macrofilariae, suggesting that, normally, electron transport is linked to ATP synthesis (Mendis and Townson 1985). A final point is that subcellular fractions prepared from whole worms oxidised α-glycerophosphate, succinate and malate, suggesting that they too possess branched electron transfer pathways, with electrons entering at the level of rhodoquinone. This topic is discussed in more detail later in the chapter (Section 2.3.2). 'Branched' electron transport systems are common in helminths and the rates of substrate oxidation in *B. pahangi* were comparable with those reported for other nematode parasites. Finally, as yet another complicating factor, submitochondrial particles from *B. pahangi* and *D. viteae* possess a fumarate reductase system which may be important in energy metabolism under aerobic conditions (Fry and Brazeley 1985). Taken together, these observations imply that other pathways of energy metabolism exist in addition to simple homolactate fermentation.

In a study on another filarial worm, *Litomosoides carinii,* the uptake of glucose during *in vitro* incubation is significantly higher in the presence of oxygen (Ramp and Köhler 1984; Ramp, Bachmann and Köhler 1985). Under anoxic conditions, glucose is almost quantitatively converted to lactic acid. When, however, oxygen was admitted to the system appreciable quantities of acetate, acetoin (acetylmethylcarbinol) and carbon dioxide were formed as well as lactate.

Simultaneous formation of acetate and acetoin is often symptomatic of the pyruvate dehydrogenase complex (Figure 2.17). This complex comprises at least three enzymes (pyruvate dehydrogenase, lipoyl transacetylase and dihydrolipoyl dehydrogenase) that require five coenzymes (thiamine pyrophosphate, lipoate, acetyl CoA, flavin adenine dinucleotide and NAD^+). It catalyses the important series of reactions that, under aerobic conditions, prepare carbon from the glycolytic pathway for entry into the tricarboxylic acid cycle (Figure 2.1). It is presumed that much of the carbon dioxide observed during aerobic respiration in *L. carinii* is generated by the activity of this cycle.

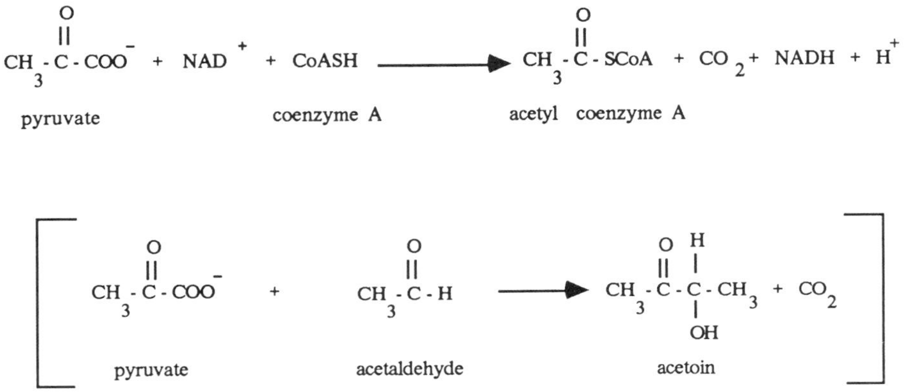

$$CH_3\text{-}\overset{\overset{\displaystyle O}{\|}}{C}\text{-}COO^- + NAD^+ + CoASH \longrightarrow CH_3\text{-}\overset{\overset{\displaystyle O}{\|}}{C}\text{-}SCoA + CO_2 + NADH + H^+$$

pyruvate   coenzyme A   acetyl coenzyme A

$$\left[\; CH_3\text{-}\overset{\overset{\displaystyle O}{\|}}{C}\text{-}COO^- + CH_3\text{-}\overset{\overset{\displaystyle O}{\|}}{C}\text{-}H \longrightarrow CH_3\text{-}\overset{\overset{\displaystyle O}{\|}}{C}\text{-}\overset{\overset{\displaystyle H}{|}}{\underset{\underset{\displaystyle OH}{|}}{C}}\text{-}CH_3 + CO_2 \;\right]$$

pyruvate   acetaldehyde   acetoin

[The condensation leading to the formation of acetoin is written within the brackets]

Figure 2.17: The Reaction Catalysed by the Pyruvate Dehydrogenase Complex

All the enzymes of glycolysis are present in *L. carinii*. They occur at much higher activities (except phosphofructokinase) than found in rat liver. *L. carinii* also possesses the tricarboxylic acid cycle enzymes. As in mammals, they are situated in the mitochondrion and, when compared on the basis of mitochondrial protein, their specific activities are similar to those from rat liver. Studies with ^{14}C-labelled substrates show unequivocally that adult *L. carinii* is capable of oxidising glucose completely to carbon dioxide and water, using molecular oxygen as the terminal electron acceptor. Further, inhibition studies suggest that there is a functional aerobic electron transport system. Overall, the results suggest that 2% of utilised carbohydrate may, in normal respiration *in vitro*, undergo complete oxidation to carbon dioxide and water. Starting from 100 moles of glucose, the energy yield of 98 moles converted to lactate is 196 moles of ATP. The potential energy yield of 2 moles of glucose converted to carbon dioxide and water by the tricarboxylic acid cycle is 72. From this, it is clear that, under the conditions of the experiments, 27% of energy generation in *L. carinii* could be aerobic.

The malate dismutation

The second type of metabolism is found in, and has been most extensively studied in *Ascaris* spp., *Hymenolepis diminuta* and *Fasciola hepatica* . *Ascaris* -like metabolism is a special case and is discussed later in this section. Carbon from glucose follows the glycolytic route to the level of phosphoenolpyruvate (PEP). A subsequent carbon dioxide fixation step yields oxaloacetate which is, in turn, reduced to malate. Malate then enters the mitochondrion where it is metabolised further.

In this type of metabolism the key reactions effectively achieve a dismutation of malate. In a dismutation system there are two routes for the metabolism of a given compound, of which one is oxidative and the other is reductive. The oxidative step drives the reductive step. In the case of the malate dismutation, the additional need for the mitochondrion to remain in redox balance requires one molecule of malate to be oxidised while two molecules are reduced (Figure 2.18). This is because the

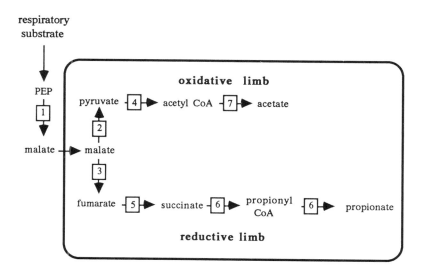

Reactions written inside the box take place in the mitochondrion.

1, phosphoenolpyruvate carboxykinase and malate dehydrogenase.
2, 'malic' enzyme. 3, fumarase. 4, the pyruvate dehydrogenase complex.
5, the fumarate reductase complex. 6, the succinate decarboxylase system.
7, various transferases or thiokinases.

Figure 2.18: The Malate Dismutation in Parasitic Helminths

oxidative arm of the branched pathway possesses two reactions that each generate one reducing equivalent per molecule of malate oxidised, while that of the reductive arm has only a single reaction that utilises reducing equivalents. The products, usually acetate and succinate or propionate, are then excreted as the free acids.

The malate dismutation, unlike homolactate fermentation, involves two subcellular compartments – the cytosol and the mitochondrion. For continued functioning both compartments must remain in redox balance. It will immediately be apparent that lactate and malate production are equivalent in metabolic terms. They are produced in the cytosol and malate dehydrogenase has the same role in regenerating NAD^+ as lactate dehydrogenase. Both pathways are also equivalent energetically because phosphoenolpyruvate carboxykinase has a similar role to pyruvate kinase as it brings about the synthesis of ATP. It does it, however, at one remove because its preferred cofactor is guanosine diphosphate (GDP) not ADP, and it generates GTP. An additional reaction is thus necessary for conversion to ATP.

Phosphoenolpyruvate carboxykinase (PEPCK) is an important enzyme. It catalyses the following reaction:

$$\text{phosphoenolpyruvate} + \text{GDP} + P_i + CO_2 \ \text{---}\!\longrightarrow\ \text{oxaloacetate} + \text{GTP}$$

In higher animals, it is usually encountered in gluconeogenesis (the synthesis of glucose from non-carbohydrate precursors) in the reverse reaction. It is worth noting that there is, in the parasitological literature, no convincing demonstration of gluconeogenesis in parasites except from intermediates at the level of the triose phosphates.

It appears that in parasites and many invertebrates PEPCK has a catabolic function whereas in the vertebrates its role is primary anabolic. Other examples of similar enzymes with different metabolic roles in parasites include certain reactions of the tricarboxylic acid cycle (malate dehydrogenase and fumarase) and the reversed β-oxidation-like reactions in *Ascaris*, as discussed below.

Although malate and lactate are metabolically equivalent, their subsequent fates are different. Lactate is excreted as lactic acid while malate enters the mitochondrion and undergoes the dismutation described above. The result is the net generation of more ATP than can be achieved by homolactate fermentation (see Table 2.1).

This simple picture is not the whole truth. It is complicated by recent experiments demonstrating that pyruvate, as well as malate, may act as a mitochondrial substrate. It is also complicated by the superimposition of varying degrees of aerobic metabolism, involving tricarboxylic acid cycle activity. Further, respiratory patterns may change during development. For example, the young liver fluke is an almost totally aerobic organism with tricarboxylic acid cycle activity. By 24 days post-infection, this activity only accounts for less than 20% of total respiration, the bulk of it being taken over by aerobic acetate formation. The latter peaks at about 50 days, and then declines as the adult pattern – the malate dismutation – takes over and accounts for 90% of respiration after about 114 days (see Figure 1.4).

A variant of the malate dismutation has been proposed for some of the filarial worms. It is interesting that phosphoenolpyruvate carboxykinase and malic enzyme activities have been found in *Dirofilaria immitis,* which is considered to be a homolactate fermenter (Brazier and Jaffe 1973). This, together with the absence of evidence for any of the pathways leading to the excretion of succinate, propionate or acetate, has led to the suggestion (McNeill and Hutchinson 1972) that pyruvate might be derived, at least in part, from the cleavage of malate which would give a pathway reminiscent of the oxidative arm of the dismutation (Figure 2.19).

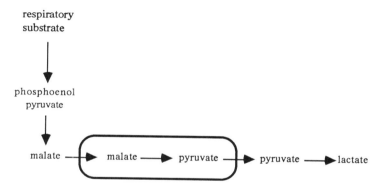

Reactions written inside the box take place in the mitochondrion.

Figure 2.19: A Possible Pathway to Lactate in Filarial Worms

It is clear from experimental work with the cestode *Hymenolepis diminuta,* which is not a homolactate fermenter, that mitochondrial malate contributes substantially to the cytosolic pyruvate pool (Behm, Bryant and Jones 1987). This demonstrates that there is considerably more communication between mitochondrial and cytosolic compartments than allowed for by the simplified metabolic schemes described in the literature.

However, as Barrett (1983) remarks, there is no experimental evidence to support the scheme proposed for filarial worms, and other functions for the enzymes, such as NADPH generation for synthetic reactions and transport of acetyl CoA across mitochondrial membranes, may be more important.

A more complex metabolic scheme has been described in *Ascaris* spp., and probably occurs in a number of other nematode genera. It is doubtful whether it occurs amongst the Platyhelminthes or the Acanthocephala. This scheme is an

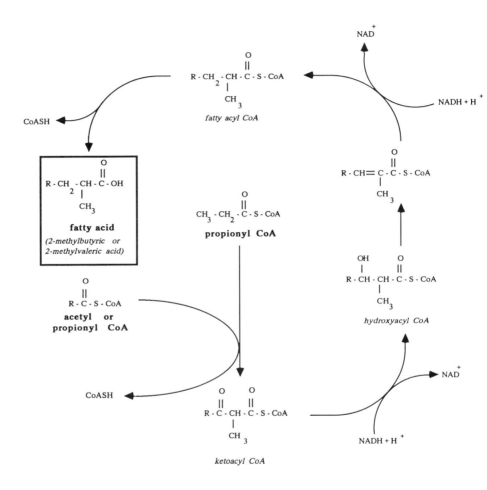

CoASH, free coenzyme A.

Propionyl coenzyme A initially condenses with either acetyl CoA or another molecule of propiony CoA. With acetyl CoA , the end product is 2-methylbutyric acid, and an intermediate   is 2-methyl crotonyl CoA (tiglyl CoA). Tiglic acid was early detected in studies of nematode metabolism. With propionyl CoA, the end product is 2-methyl valeric acid. The relationship to the β-oxidation pathway for fatty acids is clear when this figure and Figure 2.21 are compared.

Figure 2.20: Probable Route of Synthesis of Excretory Fatty Acids in *Ascaris*

elaboration of the malate dismutation, extending it by a number of redox and condensation reactions.

In *Ascaris* the main end products of glucose catabolism are 2-methylvalerate and 2-methylbutyrate, with small amounts of propionate, acetate and other volatile fatty acids (Figure 2.20). The work of Komuniecki, Komuniecki and Saz (1981a,b) suggests that there are two possible, related processes involved. One is initiated by a condensation of acetyl CoA with propionyl CoA, the other by the condensation of two molecules of propionyl CoA. The important features of these pathways are that they involve (i) two reductive steps, in which NADH is consumed, (ii) an acyl CoA intermediate, and (iii) the regeneration of CoA at the end of the reaction sequences. They are thus remarkably similar to a reversed β-Oxidation pathway (Figure 2.21).

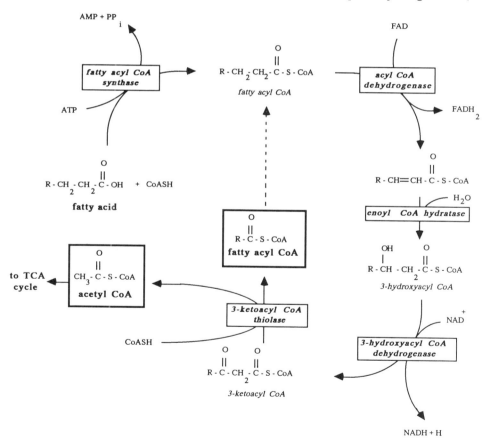

CoASH, free coenzyme A; TCA, tricarboxylic acid cycle; FAD, flavin adenine dinucleotide

Figure 2.21: β-Oxidation of Fatty Acids

β-Oxidation of fatty acids occurs in higher organisms and is remarkably efficient at generating ATP. It is conspicuously absent from adult helminths, which have only a very limited capacity for metabolising fats, although it occurs in some larval forms.

2.3.2 *Electron transport in helminths*

In parasitic helminths it is now widely accepted that, while many of the adults possess relatively low activities of aerobic electron transport systems that are comparable in function with those of obligate aerobes, their specialised electron transport systems are anaerobic.

The idea that aerobic and anaerobic electron transport systems can exist in the same organism is a challenging one to experimentalists, because of the great difficulty of assigning function to the components of the system. All parasitic helminths and Protozoa show, *in vitro*, a measurable uptake of oxygen; while some of the oxygen is no doubt used for synthetic reactions, a part of it is used in respiration. It has been noted earlier that only a small oxygen consumption may, because of the greater energetic yield of aerobic respiration, contribute greatly to the overall energy economy of the organism, but this notion presupposes the existence of a mammalian type of electron transport system. This is certainly not so in many protozoans where there appears to be no energy conservation directly associated with respiration.

Barrett (1981) poses three questions that must be answered before the importance of oxygen in helminth respiration can be assessed. They are (i), what is the mechanism of oxygen uptake? (ii), does ATP formation accompany oxygen utilisation? and (iii), how much ATP is formed?

The first question is not easy to answer. Almost all of the components of anaerobic and aerobic electron transport systems will react with oxygen if given the chance. Such a reaction may have little to do with biological function but may simply be a result of their electrochemical characteristics. That this is certainly part of the answer is indicated by a number of well-authenticated phenomena. The first is that the consumption of oxygen by many helminths depends on the partial pressure of the oxygen available. Such parasites are respiratory 'conformers', which usually indicates that it is the flavin component of the electron transport system that is responsible for the oxygen uptake. However, the flavin oxidases have a low affinity for oxygen which probably disqualifies them from the role of terminal oxidase when oxygen concentrations are low.

Further, aerobic respiration in helminths and many Protozoa is not particularly sensitive to inhibition by cyanide. This is a property shared by many free-living representatives of these groups – it was noted as early as 1919 by Hyman – and implies, because of the high affinity of cyanide for the classical cytochrome oxidase, that the aerobic oxidase is present in relatively low activities.

Finally, it has been observed that hydrogen peroxide accumulates during oxygen consumption by many groups. Hydrogen is not a product of the reaction of cytochrome oxidase with oxygen as it catalyses the sequential transfer of four electrons to oxygen in a tightly coordinated manner so that the formation of toxic intermediates is minimised. This is not a feature of other terminal oxidases.

Although the above observations demonstrate electron transport by systems that do not include cytochrome oxidase, they do not necessarily permit us to conclude that cytochrome oxidase activity is absent from helminths. Indeed, Cheah (1972) has presented evidence for the existence of aerobic electron transport systems, similar to those of the mammals, in large cestodes and nematodes (Figure 2.22). However, the concentrations of the aerobic components are much lower than those of the anaerobic components.

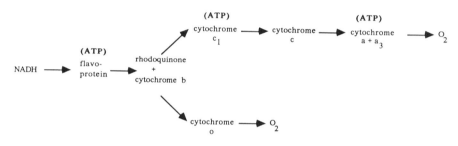

Figure 2.22: Aerobic Electron Transport in Parasitic Helminths (after Cheah 1972)

A further complication is that the electron transport systems may be branched, and therefore possess several terminal oxidases (Figure 2.23). In *Ascaris*, for example, there appears to be two branches; one leads to cytochrome oxidase, as in mammals, but a branch at about the level of cytochrome b leads to a b-type cytochrome that has been designated cytochrome o. Cytochrome o is autoxidisable and widely distributed in microorganisms. The product of its reaction with oxygen is hydrogen peroxide, not water, so it is thought most likely that cytochrome o is acting as the terminal oxidase when helminths consume oxygen.

In aerobic respiration in helminths, then, electrons are transported along a chain of electron carriers either to cytochrome oxidase or to alternative oxidases. Oxygen acts as the electron sink and the products of respiration are either water or hydrogen peroxide. It is presumed that either catalase or peroxidase, present in helminths in high activities, renders the potentially dangerous hydrogen peroxide harmless. The path to water, presumably, by analogy with the mammalian system, possesses three proton translocating sites from which ATP can be synthesised. And that is the uncertain answer to Barrett's second question. As for the third question – how much? – there is no satisfactory answer yet.

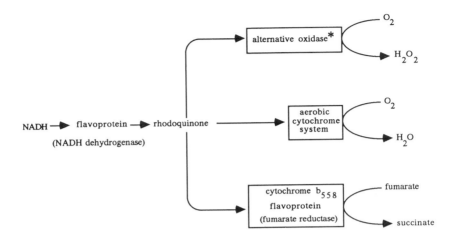

* the "alternative oxidase" may be the same as cytochrome b$_{558}$ or it may be a different form of the same molecule.

Figure 2.23: Branched Electron Transport in Parasitic Helminths

There are two other important differences between helminth and mammalian systems. The first is in the nature of the quinone electron carrier in the chain. Instead of the common ubiquinone, many helminths capable of anaerobic, electron transport-mediated respiration possess rhodoquinone. The presence of rhodoquinone in the electron transport pathways of helminths is an extremely interesting anaerobic adaptation. The redox potential of rhodoquinone (E_0) is -63 mV, considerably more negative than ubiquinone (+113 mV) and the fumarate/succinate couple (+33 mV) (Köhler 1985). This property strongly favours electron transport in the direction of fumarate. The second difference is the possession of an enzyme complex called fumarate reductase.

The fumarate reductase system is the key to the malate dismutation. Fumarate reductases are known from microorganisms. They are distinct enzymes – distinct, that is, from the enzyme that catalyses the reverse reaction in aerobic organisms, succinate dehydrogenase. Kita, Takamiya, Furushima, Ma and Oya (1988) and Oya and Kita (1989) have reported the isolation and purification of respiratory chain complex II (succinate-coenzyme Q reductase) from *Ascaris* muscle mitochondria. It apparently contains four major and two minor polypeptides, two of which have the same molecular weight as the two subunits of mammalian succinate dehydrogenase. It also contains large amounts of a cytochrome, b$_{558}$.

Cytochrome b$_{558}$ is the major constituent in *Ascaris* mitochondria. It is reduced by NADH and oxidised by fumarate in the presence of a mitochondrial preparation, and it seems likely that anaerobic electron transport is as illustrated in Figure 2.23. In this figure the two aerobic pathways are also included. It shows a single phosphorylation site if the anaerobic route is followed, and the reactions of the anaerobic route are presumably reversible, in which case the system would behave like a succinate dehydrogenase.

Figure 2.23 also implies that the pathways are indeed branched. This remains to be confirmed. We (Behm and Bryant 1976) have pointed out that a very efficient system of recycling end products can be achieved if the 'aerobic' mitochondria and the 'anaerobic' ones are in separate compartments, cells or tissues within the worm (Figure 2.24). Examination of sections of helminths by electron microscopy shows that mitochondria within a single organism are very heterogeneous and that, in many cases, those of the more superficial tissues tend to be more cristate than those of the deeper ones. Generally, the more cristae a mitochondrion possesses, the greater its aerobic, oxidative capacity.

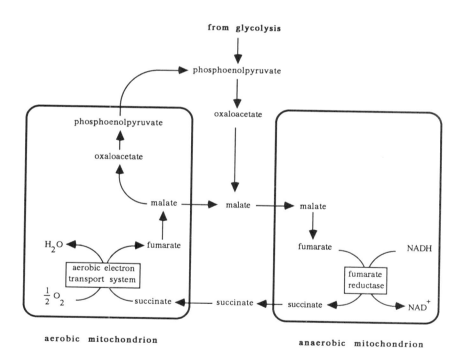

Figure 2.24: A Possible Scheme for Substrate and Product Shuttling between Aerobic and Anaerobic Mitochondria

In this scheme substrate shuttling could be achieved as follows. In the superficial tissues, the mitochondrial substrate is succinate. Succinate is oxidised to malate via a mammalian type of electron transport system and fumarase. As the tricarboxylic acid cycle is present only in low activities at best, malate accumulates and passes out of the mitochondrion. Malate diffuses or is transported to anaerobic mitochondria, for which it is the substrate. It enters and is converted to fumarate and to succinate by the dismutation reactions. There is some evidence, obtained from *Moniezia*, whose major respiratory end product is succinate, that such shuttling is possible.

If a substrate shuttle between aerobic and anaerobic mitochondria in the same organism does, in fact, exist, it would provide an excellent example of a biochemical adaptation, not necessarily to a parasitic habit but certainly to an environment with low or fluctuating oxygen tensions. The advantage to the animal

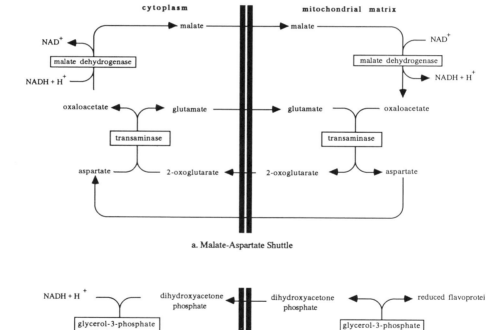

Figure 2.25: Two Intracellular Shuttles that Probably Contribute to Energy Metabolism in Parasitic Helminths

possessing such a shuttle is that it would be able to maintain the synthesis of ATP in the absence of oxygen and, as oxygen became available, to partition substrate between the two sorts of mitochondria, thus improving the efficiency of energy generation.

A major problem with shuttling systems, however, is not, as an observer might think, to show that the components of a shuttle are present but to show unequivocally that it operates *in vivo*. This is especially true for two shuttles that probably operate between the cytoplasmic and mitochondrial compartments of the cell.

One of the unrealised aims of students of energy metabolism in parasites is to draw up energy balance sheets in which the oxidation of substrate is matched to the production of ATP. Under ideal conditions, it ought to be possible to account for all ATP synthesised in terms of substrate utilised or, to put it another way, in terms of reducing equivalents ($NADH + H^+$ and reduced flavoprotein) generated. If a balanced budget can be drawn up, it implies that there are no unknown processes contributing to respiratory metabolism. Unfortunately, almost all budgets fail to balance. The number of reducing equivalents generated from known substrates are usually insufficient to account for the amount of ATP generated. Some other process or processes are obviously at work.

Two such processes are illustrated in Figure 2.25. They are the malate/aspartate and the glycerol phosphate shuttles. Kohlhagen (1988) has shown that all the components of the shuttles occur in a variety of parasitic helminths, and has provided circumstantial evidence that they do indeed operate. As the figure shows, their activities would serve to increase the number of reducing equivalents within mitochondria at the expense of reducing equivalents in the cytosol. These additional reducing equivalents, over and above those that can be derived from commonly used substrates such as glucose, probably derive from amino acid and fat metabolism. Oxidative metabolism in parasites is thus a far more complex process than *in vitro* studies with single substrates tend to indicate.

2.4 Adaptation and energy metabolism

Barrett (1989) argues that the anaerobic metabolism of parasites differs from that of free-living invertebrates in that it is continuous, not transient, and persists in the presence of oxygen. In addition, it is variable; it varies between and within species. Quite clearly, constraints other than simple metabolic efficiency operate differently in different species and populations. Variability in a system that apparently, on first principles, ought to be conservative has confounded many parasite biochemists. Barrett (1989) stresses the concept of the 'power' of a metabolic pathway – the notion that the rate at which a pathway generates ATP is as important as the amount of ATP/mole glucose generated. In Barrett's words "the end products selected by helminths must represent a compromise between substrate conservation and rate of working". 'Power' is a concept that rehabilitates a view once held that parasites 'living in the midst of plenty' are under no pressure to maximise the oxidative efficiency of their energy-generating pathways. For example, the energy output during production of lactate from glucose, though less oxidatively efficient than catabolism to CO_2 and H_2O may be adequate provided that the enzymes in the glycolytic pathway can process large amounts of (unlimited) substrate. Further, organisms that carry out mixed fermentations in various subcellular compartments have more opportunity for maintaining appropriate redox levels in those compartments, by altering carbon flow. This may allow them to adapt to changing circumstances in their immediate environments, and are different for every host-parasite system.

There are thus two qualitatively different ways in which organisms may vary. They may vary because the starting materials were different. Extreme examples may be found in the parasitic Protozoa. *Tritrichomonas* does not possess mitochondria; it has hydrogenosomes. Müller (1976) has speculated that the hydrogenosomes may have been derived from a symbiosis with a *Clostridium*-like anaerobe. *Entamoeba* contains no mitochondria at all. Of course, alternative explanations are possible – these protozoans once had 'normal' mitochondria, lost them and acquired other systems to carry out their function. This does not seem likely, and is certainly not the most parsimonious explanation; however, there is no reason to suppose that evolution is parsimonious.

It is possible, then, that parasites are different because they derive from a different symbiotic event from other organisms; their starting materials were different. But the second way in which they might have acquired their differences from free-living animals is during their evolution after the primitive symbiotic event. One of the major determinants in their subsequent evolution appears to have been their ability to colonise environments with low or fluctuating oxygen tensions.

Elsewhere (Bryant 1989) it has been argued that the possession of an anaerobic metabolism in certain of the lower Metazoa was an essential prerequisite for the assumption of the parasitic habit by helminths. In retrospect, this is an over-simplification because it does not explain the acquisition of aerobic systems of energy metabolism by many helminth groups. Many helminth larvae have aerobic metabolism, a complete tricarboxylic acid cycle and the ability to oxidise fatty acids; the adults that may possess none of these pathways must certainly carry the genetic information. Neither does this argument take into account modern ideas that the assumption of the parasitic habit in the Nematoda occurred not once but many times, over a wide time span. However, it does explain the similarities between many free-living anaerobic Metazoa and parasitic helminths.

The environments occupied by free-living anaerobes are low or deficient in oxygen, rich in carbon dioxide, and have a pH around 7 and a redox potential that facilitates electron transfer to a cytochrome of the b group. These are exactly the conditions that are encountered by helminths in the vertebrate intestine. Given these similarities, one is forced inexorably to the conclusion that the energy metabolism of parasites has nothing to do with the parasitic habit. Instead, it is an adaptation to the *habitats* of parasites, habitats that, in their important evolutionary determinants, are simply special cases of a class of habitats occurring widely in nature.

If low oxygen availability is an important selecting factor for helminth parasites, the size of a parasite may also be important in determining the type of energy metabolism. Fry and Jenkins (1984) found a good inverse correlation between the thickness of the body and the aerobic capacity of nematodes. In tapeworms, cristate mitochondria are found more frequently in the superficial tissues, suggesting that there is a decreasing oxygen gradient from the surface to the centre of the worm. This may also provide an explanation for the decline in the contribution of aerobic pathways to metabolism during development. As worms increase in size, poor diffusion of oxygen into the deeper tissues of large parasites may preclude complete dependence on oxygen as an electron acceptor (Köhler 1985).

One of the more obvious characteristics of energy metabolism in parasitic helminths and Protozoa is the production and excretion of a range of relatively highly reduced end products. Prominent among them are the various acids, and it is conventional wisdom to attribute their production to the inevitable consequences of anaerobic metabolism. Is it possible that there are other facets of acid excretion that make this particular option an energetically favourable strategy?

In 1970, in an important paper that examined the relationships between biochemical adaptation and so-called loss of genetic capacity in parasites, Fairbairn discussed the properties of organic acids excreted by parasites. He observed that, as the dissociation constants for lactic and succinic acids were high ($K_a = 13.87 \times 10^{-5}$ and 6.63×10^{-5}, respectively), they would be less likely to dissociate at physiological pH and would therefore pass more easily through a lipid membrane than the dissociated ionic species (protons, and anions of acids). In other words, Fairbairn considered that there was an energetic advantage in excreting certain acids. This, however, does not explain why so many parasites excrete acetic acid, which has a much lower dissociation constant (1.8×10^{-5}) or why they take a step beyond succinic acid and produce propionic acid, which has an even lower dissociation constant of 1.34×10^{-5}.

A second explanation was offered by Bowlus and Somero (1979). They considered that the properties of succinic acid were such that it was compatible with the enzyme systems that produced it. In defence of this point of view, they argued that succinate actually stabilises protein structure. This is no doubt true, and offers a very nice problem in coevolution. Did the need to produce succinate shape the metabolic system, or did the system dictate the end product? It comes dangerously close to the Panglossian view, that 'all is for the best in this best of all possible worlds', of the perfect adaptationists. The reality is certainly one of compromise between different evolutionary constraints.

However, behind both of these suggestions is the idea that something other than the necessity to get rid of a useless molecule is needed to account for the ubiquity of acid excretion in parasites. An extremely interesting idea derives from work with micro-organisms (Michels *et al.* 1979) and has recently been reviewed by Konings (1985).

It depends on an extension, to fermentative bacteria, of the chemiosmotic hypothesis for energy generation. Bacteria may be able to create a proton motive force by the excretion of end products of metabolism across the cytoplasmic membrane. The 'energy recycling model', as it is called by Michels *et al.* (1979), is effectively the reversal of the process of solute uptake by a secondary transport system. Secondary transport proteins are set in the cytoplasmic membrane and are symmetrical, so that molecules can be translocated in both directions. The direction

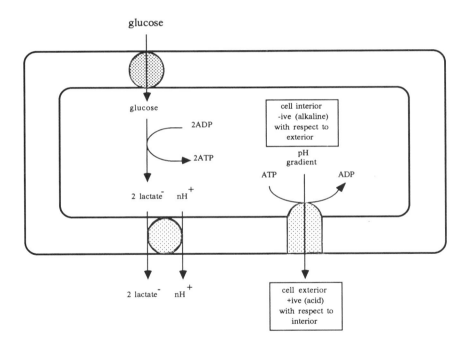

Metabolic energy is produced by glycolysis and substrate level phosphorylation. Lactate is secreted and is accompanied to the exterior by more than the strictly stoichiometrical number of protons. This has the effect of rendering the outside of the cell acid with respect to the inside. Less ATP has to be hydrolysed to maintain a proton motive force. Lactate secretion thus effectively spares ATP for bacteria.

Figure 2.26: The Bacterial Energy Recycling Model (Konings 1985)

depends on the direction of the driving force. If the driving force is a proton motive force, it will have two components; one is an electrical parameter, the membrane potential, and the other is a chemical one, the difference on each side of the membrane, set up by proton pumping.

The difference between uptake and excretion of a solute is as follows. During uptake, the energy of the electrochemical proton gradient is converted into the energy of the solute gradient. In the efflux of a respiratory end product, the energy associated with the gradient is converted into the energy of electrochemical proton gradients. These proton gradients may then be used to synthesise ATP. This is illustrated in Figure 2.26 for a bacterial homolactate fermenter.

Inhibitor and membrane vesicle studies have shown that this is a likely process in *Streptococcus cremoris* and Konings (1985) suggests that it is probable that, in other fermentative bacteria, the efflux of other end products is coupled to the generation of a proton motive force.

Of the common acids excreted by parasites, three (acetic, succinic and propionic) are of mitochondrial origin. It is an interesting speculation that the translocation of these acids to the cytosol of the parasite may set up a proton motive force that either results in the synthesis of ATP directly or exerts a sparing effect on the use of ATP that has been synthesised in the more 'conventional' way. Lactic acid is a cytosolic product and its excretion across the cell membrane into the unstirred layer at the external surface of the cuticle may also permit the setting up of a proton motive force. If there is any merit in these speculations, they would provide a fine example of biochemical adaptation to environments that are characterised by low or fluctuating oxygen tensions.

It is interesting in this context that, in *Leishmania* promastigotes, uptake of L-proline and D-glucose is driven by the proton electrochemical gradient across the cell membrane (Zilberstein and Dwyer 1985). Proton uptake is accompanied by D-glucose or L-proline uptake via a symport mechanism. This type of uptake occurs in bacteria and fungi and is probably common in Protozoa, but has not been reported in Metazoa.

3

Digestion and Uptake of Nutrients

3.1 Absorption from the medium

3.1.1 Protozoa

The barrier between the organism and its environment in all non-cyst stages of parasitic protozoa is the plasma membrane of the cell. In most cases, the external surface of the lipid bilayer is partially or completely surrounded by a 'glycocalyx' or carbohydrate-bearing external layer of molecules, such as glycoproteins or glycolipids, that form an outer coating. The thickness and composition of this carbohydrate-rich component varies with species and stages of the life cycle. In many species the presence of sialic acid residues on the oligosaccharides gives the outer face of the membrane an overall negative charge. This outer region of the membrane is the major interface between the parasite and its external environment. It may have a protective function. It is also the site of many interactions between cells or between molecules and cells. The most extreme development of a coat with a protective function is the VSG (variant surface glycoprotein) coat of bloodstream African trypanosomes which protects the infection against the host's immune attack. In order to traverse the lipid bilayer all nutrient and other molecules must also pass through or between the external carbohydrate extensions of the membrane. Thus, the uptake and permeability properties of the plasma membrane in parasitic Protozoa, as in most cells, is determined not only by the properties of the lipid bilayer and its protein components but also by the permeability properties and ionic and other interactions of the surrounding carbohydrate-enriched coat.

The lipid bilayer of all undamaged membranes is impermeable to charged molecules and to most hydrophilic molecules, with the exception of small molecules such as H_2O, O_2, CO_2 and the uncharged forms of low molecular weight acids and bases, such as ammonia or acetic acid. Lipid-soluble (hydrophobic) molecules, if they are uncharged, are able to pass freely into the bilayer and thence partition into the cytoplasm. Special mechanisms are required to effect the passage of large or charged molecules across the membrane. These mechanisms include (i) primary digestive processes that occur at or near the cell surface to reduce the size of the molecule or change its reactivity (e.g. reactions catalysed by peptidases, glycosidases, nucleases, disaccharidases, esterases); (ii) removal of the charge, catalysed by enzymes such as phosphatases and nucleotidases, to liberate an uncharged species that can traverse the bilayer, and (iii) the presence of highly stereospecific transport proteins (carriers) in the bilayer that form a gated pore or catalyse solute exchanges or net uptake reactions. In general, the specific uptake properties of any membrane are determined by the nature of the carriers present in the membrane and this is determined genetically and developmentally for each type of cell.

Parasitic Protozoa are no exception to these principles and most research effort on nutrient uptake in this group has been directed towards determining the precise properties of the uptake processes in these organisms. Clearly, the types of carrier present will be a reflection of the nutrients available to or actually utilised by the parasite. It is also likely that the kinetic properties of the carrier would reflect the parasite's ability to compete with host cells for uptake of any particular solute. We will briefly discuss what is known about uptake mechanisms in the major protozoan groups, paying particular attention to uptake processes for glucose, the substrate for energy metabolism in most of these groups. Glucose uptake is especially important for kinetoplastids and plasmodia as they store no carbohydrate reserves.

An additional complication presents itself to researchers interested in investigating the uptake of nutrients by certain groups of parasitic Protozoa. These groups are intracellular parasites at certain stages of the life cycle. It is possible to measure the uptake properties of parasites liberated from their host cells, but the process of liberation invariably causes some membrane damage that affects permeability, so the results from this type of study must always be interpreted with caution. It is also possible, and perhaps more valid, to perform uptake studies with the parasites inside their host cell. It is known that the permeability properties of host cells may change as a result of infection and, with some ingenuity and a sound knowledge of the properties of the (normal) host cell, much may be determined about uptake processes in the parasite itself as well as its nutritional relationship with the host cell.

The bloodstream forms of *T. brucei* and related parasites possess a specific carrier for glucose that has different properties from the mammalian or bacterial carrier. It has a high affinity for glucose, with a K_m of 4.03 mM compared with 80-90 mM for the mammalian carrier, although the parasite V_m is lower (Game, Holman and Eisenthal 1986). This high affinity for glucose permits the very high glycolytic rate observed in these parasites and it has been shown in *T. brucei* that glucose uptake is in fact rate-limiting for aerobic glycolysis. Threonine uptake has been shown to be an active process in these organisms and there is also a carrier for ethanolamine. A 3'-nucleotidase, a 3'-endoribonuclease and an α-glucosidase have been identified as digestive enzymes associated with the plasma membrane. An acid phosphatase is located on the external surface of the plasma membrane in *T. cruzi*.

Transport systems for hexoses and a variety of amino acids have been identified in *Leishmania* spp. Although lipid inclusions are present in the cells, the uptake systems for lipids and their precursors have not been described. The uptake mechanism for glucose in *Leishmania* is especially interesting as it has been shown to be catalysed by a carrier that transports glucose in symport with protons (Zilberstein and Dwyer 1985). The energy for this process comes from the proton gradient across the plasma membrane and is analogous to well-known uptake systems present in bacteria and certain fungi. The proton gradient also provides the energy for uptake of proline, methionine, alanine and α-aminoisobutyrate, indicating that proton gradient-driven uptake of nutrients is an important process in this group. The proton gradient is maintained by a plasma membrane H^+-ATPase. There is some evidence for proton gradient-driven transport in *Crithidia* spp, *T. cruzi* and possibly also the unrelated malarial parasite, *P. chabaudi*. It will be interesting to see whether proton gradient-driven uptake is a common feature of Protozoa.

A number of digestive enzymes have been identified associated with the plasma membrane of *Leishmania* spp, including acid phosphatase, 3'- and 5'-nucleotidases and a major protease in promastigotes that may have both a digestive and protective function. An interesting insight into the relationship between *Leishmania* and its host macrophages has been gained by the observation that the parasites, which are unable to synthesise haem *de novo*, require haem supplementation of their incubation medium. This shows that the parasites do not utilise the haem synthesised by their host cells (Chang and Chang 1985).

Uptake processes in *Plasmodium* species have received considerable attention but have been limited by methodological problems associated with releasing intact intraerythrocytic stages from their host cells *in vitro* (see Sherman 1983). The malaria parasite is able to modify the uptake properties of its host cell. During the course of infection, the permeability of the red blood cell increases towards certain amino acids, polyols, anions and cations, especially at the trophozoite stage (Elford 1986). Indeed, this increase in permeability is a prerequisite for development of the parasite beyond the ring form. The permeability changes observed are selective and parasite species-specific. For example, in erythrocytes infected with *P. falciparum* uptake of glutamine increases but not that of isoleucine, arginine or glycine; this change does not occur in murine *P. chabaudi*.

Permeability increases in the erythrocyte membrane have been correlated, in *P. falciparum,* with structural changes in the membrane. Pores of 0.7 nm radius appear in the membrane during the course of the infection (Ginsburg, Kutner, Krugliak and Cabantchik 1985). The pores are lined with residues bearing a positive charge and are large enough to permit the passage of uncharged or negatively-charged molecules up to about the size of disaccharides. There is also a generalised permeability increase in the host cell membrane that is probably associated with the depletion of cholesterol and consequent increase in fluidity and osmotic fragility of the membrane (see Sherman 1983). This change in membrane composition is likely to be responsible for the observed increase in diffusional entry of glucose in the cell while the facilitated (i.e. carrier-supported) component of glucose uptake remains unchanged. It is interesting that these parasites appear to have no active transport of glucose, especially as they store no carbohydrate reserves.

Many of the amino acids required by intraerythrocytic plasmodia are obtained from the host cell by digestion of haemoglobin. Host cell cytoplasm is taken up by phagocytosis via the cytostome or by micropinocytosis and the protein component of haemoglobin is digested in the food vacuoles by highly active globinases. Haemoglobin is deficient in certain amino acids, such as methionine and cysteine, which are acquired from the host's plasma.

Hypoxanthine supplied by the erythrocyte is the major substrate for purine salvage in plasmodia, whereas the host cell salvages plasma adenosine for the same purpose. In this way the parasites do not compete for purine substrates directly with their host cells as hypoxanthine is a normal by-product of adenine nucleotide catabolism in erythrocytes. Hypoxanthine transport in erythrocytes is unaffected by infection.

The trophozoites of *Entamoeba* spp. are characterised by extensive internalisation of their surface membrane, as would be expected in phagocytic amoebae. This complicates uptake studies as pinocytosis appears to be responsible for uptake of certain nutrients, such as the vitamins riboflavin or niacin, whereas carried-mediated uptake (with perhaps a small component of pinocytosis) has been demonstrated for purines and other nutrients (see Reeves 1984a). Glucose uptake occurs by facilitated diffusion, catalysed by a high-affinity specific glucose carrier, and uptake by pinocytosis is insignificant. Galactose is transported by the same carrier. The uptake of glucose appears to be the only rate-limiting step in glycolysis in these organisms. Lipids are derived from the incubation medium but their transport mechanisms have not been described.

Entamoeba amoebae possess a large battery of hydrolytic lysosomal enzymes, including phosphatases, glycosidases, esterase, acid and thiol proteinases and RNAse, which are involved in digestion in the phagolysosomes and which probably supply a large proportion of the nutrients for these parasites under natural (i.e. non-axenic) conditions. They also possess an interesting plasma membrane Ca^{2+}-ATPase with unique properties that may have a regulatory function in controlling intracellular calcium concentrations (McLaughlin and Aley 1985).

Few studies have been carried out on uptake of nutrients by *Giardia* species. These parasites are distinguished by having no *de novo* pathways for pyrimidine synthesis and thus rely on salvage mechanisms to supply substrates for pyrimidine nucleotide synthesis. Transport of the pyrimidines uridine, cytidine and thymidine is carrier-mediated in trophozoites, with thymidine uptake by a separate carrier from the uridine-cytidine transporter (Jarroll, Hammond and Lindmark 1987). They are able to take up and incorporate lipids and phospholipids. Arachidonic and palmitic acids are taken up by a rapid, saturable process probably by direct partitioning from the medium into the external membrane (Blair and Weller 1987).

3.1.2 *Helminths*

Helminth parasites grow rapidly, produce large numbers of offspring and many of them proliferate asexually as well. In order to sustain this intense anabolic activity, it is not surprising to discover that they possess specialised mechanisms for the uptake and absorption of the many small molecules that they need. One such mechanism is the ability of cestodes to synthesise digestive enzymes which, while being retained on or within the surface that produced them, act to break down larger molecules into smaller ones which can more easily penetrate the tegument and cell membranes and be transported throughout the parasite. Typical examples are proteases and peptidases, that break down proteins and peptides respectively, the amylases that hydrolyse carbohydrates and lipases and esterases that act on fats and the breakdown products of fats. These, and sundry other enzymes like the nucleases and the phosphatases, are ubiquitously distributed among the parasitic groups – as they are amongst all organisms – and there is little profit in pursuing the details of their reaction mechanisms here. They are poorly studied in parasites and what is known is reviewed in detail elsewhere (Barrett 1981).

Digestion at the surface is a characteristic of the cestodes and the acanthocephalans, because both groups lack a gut. It may also be a feature of the trematodes, although this remains to be clarified. Surface digestion cannot be said to be a strictly parasitic adaptation because it is well established that vertebrate absorptive surfaces that are equipped with brush borders also carry an impressive range of enzymes. But as such surfaces predominate in the intestine, and as many helminths are intestinal parasites, it provides another example of the way in which parasites come into direct competition with their host. Parasite absorptive surfaces must be the kinetic equals of the host surface that surrounds them and this is also true for surface configuration. Microtriches are a feature of cestode absorption surfaces. One of their functions seems to be to increase absorption surface area and in this they are analogous to brush borders.

There are also physiological adaptations that increase the absorptive competitiveness of parasites. Contact digestion is a feature of the mammalian intestine; pancreatic enzymes become adsorbed onto the brush borders of the mucosal cells. Hydrolysis of food molecules thus occurs in close proximity to the absorptive surface, resulting in high concentrations of breakdown products in the immediate vicinity of the systems that transport them across the mucosa. This is a highly efficient process as it prevents dilution by the gut contents. It appears that several genera of cestodes are also capable of absorbing host pancreatic α-amylase. During the process, the enzyme becomes activated, by a mechanism not clearly understood and, although there are no studies that demonstrate the fact unequivocally, it presumably functions to the advantage of the parasite.

Indiscriminate absorption of host enzymes, however, can be a two-edged sword. As well as capturing enzymes that could be useful, parasites are potentially in danger from the activities of the enzymes themselves. For example, potentially dangerous enzymes are the proteases trypsin and chymotrypsin and pancreatic lipase which could damage cell surfaces. *Hymenolepis diminuta*, *in vitro*, absorbs these enzymes which then become irreversibly deactivated. It appears therefore, that cestodes possess at least two distinct mechanisms for dealing with host enzymes (Barrett 1981). There is no evidence for similar mechanisms in acanthocephalans.

The highly efficient absorptive surface of cestodes depends in large part on the glycocalyx, an excellent example of parasitic adaptation. The glycocalyx coats the cells with a complex structure composed of polysaccharides, glycoproteins and glycolipids. At physiological pH, the glycocalyx is negatively charged and acts as a polyanion. Positively charged substances are strongly bound to it, and the whole of the cestode surface thus acts as a complex and selective ion exchanger.

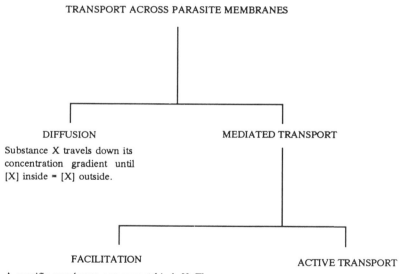

TRANSPORT ACROSS PARASITE MEMBRANES

DIFFUSION
Substance X travels down its concentration gradient until [X] inside = [X] outside.

MEDIATED TRANSPORT

FACILITATION
A specific membrane component binds X. The component/X complex diffuses down its concentration gradient across the membrane and releases X into the interior. The process stops when [X] inside = [X] outside. The rates at which X binds to the component on both sides of the membrane are then equal.

ACTIVE TRANSPORT
Specific membrane components are involved. Substance X is transported against its concentration gradient, a process which requires energy, most usually as ATP.

Figure 3.1: Types of Transport Mechanisms in Parasites and Other Organisms

By far the most complete studies of nutrient uptake that have been made are those of cestodes and, of these, *Hymenolepis* has been most intensively investigated. This work has been generally reviewed by Pappas and Read (1975) and Lumsden and Murphy (1980). In *Hymenolepis*, glucose and galactose are transported by a common system which is sodium dependent and sensitive to phlorizin. By analogy with mammalian systems, this implies that carbohydrate uptake is energy dependent and explains why tapeworms *in vitro* are capable of removing glucose from the incubation medium until its concentration falls to undetectable levels. Other carbohydrates enter by passive diffusion. The uptake of glycerol, on the other hand, shows two components – an active one, distinct from the hexose transporting mechanism, and one dependent on diffusion. The difference between active and passive transport is illustrated in Figure 3.1.

The pattern for amino acid uptake is more complex. Competition experiments have detected at least six amino acid porters that are designated according to the amino acid – or group of amino acids – for which they show the greatest affinity. They are the dicarboxylic, glycine, serine, leucine, phenylalanine and the basic amino acid porters. They are not absolutely specific, they mostly show mediated or facilitated transport, and uptake obeys simple first-order kinetics.

More complex still are the mechanisms for uptake of purines and pyrimidines, the carriers for which have both transport sites and activator sites. There are two such carriers; one for thymine and uracil, and one designated hypoxanthine 1, that also transports adenine and guanine. A third, called the hypoxanthine 2 carrier, does not have an activator site. The activator and transport sites of the first two carriers interact allosterically; binding of the purine or pyrimidine at the activator site presumably causes a change in the configuration of the transport site that facilitates the passage of the compound across the membrane (Figure 3.2). The mechanism no doubt reflects a metabolic imperative as most helminths are incapable of making their own purines and pyrimidines *de novo*.

Many other compounds are, of course, taken up by cestodes. They include the breakdown products of lipids – fatty acids, for example, are taken up by a form of mediated transport, and glycerol uptake has already been discussed – and various vitamins. Unfortunately, the mechanisms of uptake are not well studied and not enough is known about them to permit any discussion about whether they are parasitic adaptations or general properties of absorptive surfaces.

There are some interesting differences between nutrient uptake in acanthocephalans and cestodes. Hexose uptake in acanthocephalans occurs by a facilitated mechanism and not by an energy-dependent process, as it is not accompanied by sodium exchange, nor is it phlorizin sensitive. There are two sites, one for glucose and the other for fructose. Amino acids enter by a mediated system but simple diffusion also plays a part.

Nutrient uptake in trematodes is different from that of cestodes. *Fasciola hepatica* takes up hexoses by a passive mediated process, while *Fasciola gigantica* employs a mediated process that is phlorizin sensitive. It is interesting to find such a difference within a genus, but mechanisms also differ between genera. Schistosomes are variously described as taking up hexoses by diffusion and by an active phlorizin-sensitive process. The method by which the female, in the gynecophoral groove of the male, obtains glucose is particularly noteworthy. Glucose passes into the female by diffusion along a concentration gradient from the medium and through the male. In so doing, the glycogen reserves of the male are depleted (Cornford and Fitzpatrick 1985).

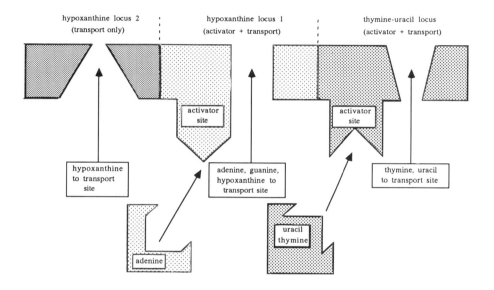

At hypoxanthine locus 1 and thymine-uracil locus, initial binding of the purines to the activator sites enhances purine transport

Figure 3.2: Purine and Pyrimidine Uptake in a Tapeworm, *Hymenolepis diminuta* (based on Pappas and Read 1975)

An interesting hypothesis has been proposed recently, by Uglem and coworkers (Uglem, Lewis and Larson 1985) that may explain the great variation in hexose uptake mechanisms in trematodes. They investigated three species of *Proterometra: P. macrostoma,* a pharyngeal ectoparasite of fish, *P. edneyi,* a stomach parasite of fish and *P. dickermani,* a tissue parasite of the snail host. The ectoparasite absorbs glucose by diffusion only, the stomach parasite employs a passive mediated process, whereas the tissue parasite displays a sodium-dependent active transport, extremely sensitive to phlorizin. It is suggested that this pattern correlates with the environmental niche occupied by the parasite and that, where the parasite is in close apposition to actively metabolising host tissues it is most likely to possess an active transport system to enable it to compete effectively for nutrients.

Little is known about absorption of other nutrients by trematodes. Mechanisms of amino acid uptake vary within the group. In *Fasciola*, it is by diffusion, whereas the schistosomes, in addition to diffusion, also possess a form of mediated transport that may be sodium sensitive. In schistosomes, as well, there appear to be at least five carriers for purine uptake. There is almost no information about lipid uptake.

Cestodes and acanthocephalans have no intestine, so that the only way in which nutrients can enter the parasites is through the body wall. Although there is ample evidence that the trematode integument is, like that of cestodes, syncytial and modified for absorption it possesses no microvilli nor microtriches. Trematodes, however, have a branched intestine with a single opening and this affords another route for digestion. Liver flukes at various times have been observed feeding on the tissues of the bile duct walls and on blood, but the individual contributions of tegumental and intestinal absorption to the overall economy of the parasite have not been extensively studied.

In this context, it is interesting to note what may be a highly specialised biochemical adaptation in the liver fluke. One of the major end products of its metabolism is the heterocyclic amino acid, proline. After the initial elucidation of the pathway by which it is produced (Figure 3.3), Isseroff and coworkers have shown, in an elegant series of experiments, that ectopic liver fluke, restrained from migration by a porous chamber and placed in the peritoneal cavity of rats, induce bile duct hyperplasia – that is, the characteristic thickening of bile ducts that is observed during the liver fluke infections. Further, they have shown that infusion of proline into the hosts causes hyperplasia even in the absence of liver flukes (Isseroff, Girard and Leve 1977; Isseroff, Sawma and Raino 1977). There are interesting differences between the proline-produced hyperplasia and that produced naturally by infections with liver fluke, so the possibility of the existence of other factors acting with proline cannot be ruled out. Even so, it is tempting to speculate that proline secretion is a specific adaptation in liver flukes that causes the proliferation of one type of host cell that the parasite can readily exploit. In other words, the presence of the parasite improves its environment. The proline pathway has probably not arisen *de novo* and is not unique to the liver fluke. It probably occurs in all parasites, and all animals. In the fluke it might seem that selection pressure has resulted in a change in the genes that are responsible for regulating the pathway. Unfortunately this hypothesis is not supported by the fact that *S. mansoni*, which does not infect the liver, also excretes proline and induces bile duct hyperplasia in mice (Bedi and Iseroff 1979).

Nematodes are bounded by a cuticle that is considered to be permeable only to water and other low molecular weight compounds, although there are reports of filarial worms absorbing nutrients by the transcuticular route (see below). The intestine of the nematode is a simple tube, one cell thick, the inner surface of which is covered with microvilli that increase the absorptive area. Little is known of absorption processes in the nematode intestine.

Adult filarial worms have a functional digestive tract. Host erythrocytes have been observed in the intestine of the dog heartworm, *Dirofilaria immitis*, although ingestion has not been demonstrated *in vitro*. In this, it is similar to *Brugia pahangi*. A number of adult filariae have been shown to absorb nutrients through the cuticle. For example, *D. immitis* takes up and metabolises D-glucose,

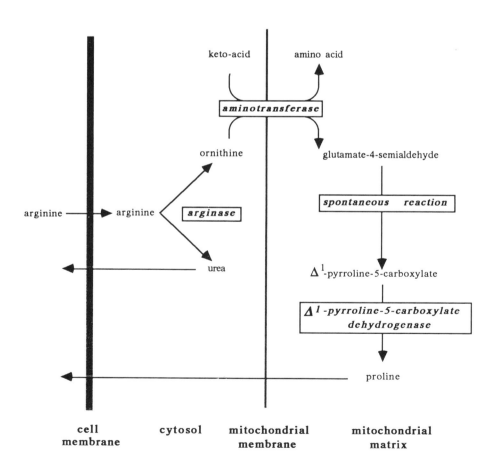

Figure 3.3: Proline synthesis in *Fasciola hepatica* (modified after Kurelec 1975)

adenosine, uridine and uracil even when the anterior portion, bearing the mouth, is immersed in a nutrient free solution while the posterior part is bathed in nutrient medium. Uptake was found to be selective, for L-glucose, sucrose and thymidine are excluded, but the details of the transport mechanisms are unknown. There are a number of enzymes present in the cuticle that, in other tissues, are associated with nutrient uptake and digestion. For example, high levels of acid protease, capable of hydrolysing haemoglobin, are present in the hypodermis of *D. immitis*.

Microfilariae of *D. immitis* consume large quantities of glucose during *in vitro* incubation, and the rate of glucose consumption is dependent on concentration. Most of the glucose is used in respiratory metabolism. They also use adenine, adenosine, uracil and uridine, incorporating it into RNA, but not thymidine. As microfilariae do not possess a functional intestine, uptake is almost certainly accomplished by the cuticular route. As in the adult, uptake is selective, because many amino acids are not absorbed from the medium during *in vitro* culture while others are.

3.2 Folic acid metabolism

Folic acid and its derivatives are extremely important in metabolism because they assist in the transfer of one-carbon groups (formyl, methyl or hydroxymethyl) to acceptor molecules during biochemical syntheses. Folic acid combines pteridine, *p*-aminobenzoate (PABA) and glutamate moieties in one complex molecule (Figure 3.4). In particular, folic acid is concerned with purine and pyrimidine metabolism. Folic acid metabolism is discussed in more detail in Chapter 5. Figure 5.2 shows the various interactions of intermediates in folic acid metabolism.

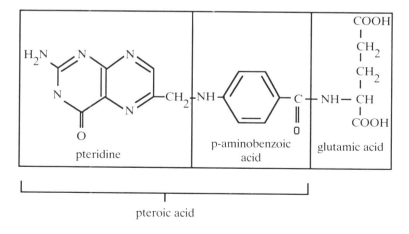

Figure 3.4: The Structure of Folic Acid

These reactions are of such fundamental importance to metazoans that it is likely that they are highly conserved and not susceptible to much variation. However, an important difference between filariae and their hosts is the ability of filariae to oxidise 5-methyltetrahydrofolate to 5,10-methylene tetrahydrofolate. This could be an adaptation to the parasitic habit because it provides them with an additional substrate for folate metabolism other than folate itself. The host is capable only of the reverse reaction.

3.3 Lipids and lipid metabolism

The metabolism of lipids and lipid-like compounds in parasites is a much neglected field of study. On the other hand a considerable amount is known about their distribution in the parasitic groups. Lipids in parasitic helminths have been reviewed extensively by Frayha and Smyth (1983) so a detailed account is not attempted here.

The term lipid encompasses a wide variety of chemical structures that include acylglycerols, fatty acids, phosphoglycerides, waxes and steroids. For the sake of simplicity here we identify three groups on the basis of their behaviour in commonly used separation systems. They are (i) glycerides – that is, esters of glycerol and fatty acids – and free fatty acids, (ii) phospholipids and (iii) unsaponifiable lipid material, which contains the waxes, terpenes and steroids. The functions of the various lipids in parasites are understood only by analogy with other, well-studied organisms. It is likely that the analogy is well founded, as parasites, like other organisms, contain cells with external and internal membranes. Many of the different types of lipids form a major part of the structure of membranes. Clearly, these lipids are not specifically parasitic adaptations, but are basic requirements for all eukaryotic cells. There are, however, a number of peculiarities about parasites that are worth noting even if, at present, we cannot explain them.

All parasitic Protozoa seem to require lipids in their growth media. The specificity of their requirements has usually not been determined and, most often, a generalised source of lipids – such as serum – suffices. Parasitic helminths, too, take up lipids, fatty acids in particular, from their incubation media. In adult helminths this is not, apparently, for the purpose of oxidation and the generation of ATP as they lack the β-oxidation pathway (Chapter 2). Neither does it seem to be for elaboration into other, specifically helminth, molecules, as they have only limited capacity for metabolising fatty acids; for example they are able to lengthen the chain and saturate certain unsaturated bonds.

Yet some important purpose must be served by the fatty acids, as parasitic helminths are characteristically rich in them. The level of free fatty acids in animal tissues is usually of the order of 1% of the total lipids. In helminth tissues, they may comprise as much as 80% or as little as 10%. The total lipids may vary from about 1% to about 35%. The major fatty acids usually contain 16 or 18 carbon atoms, and the latter are often present as the unsaturated oleic and linoleic acids. In fact, helminths have a rather high level of unsaturated fatty acids compared with tissues from free-living animals. Odd-numbered fatty acids (C15 and C17) usually comprise less than 1% of the total, except in the filarial worms where branched C17 fatty acids may make up 8%.

The composition of the fatty acids found in parasitic helminths is often a function of the diet of the host – in cestodes particularly the relative proportions of fatty acids can be changed by changing host diets. They also reflect the constituent fatty acids of the host; shark tapeworms contain C20 and C22 polyunsaturated fatty acids, for example (Buteau, Simmons and Fairbairn 1969). In sharks, the possession of long-chain unsaturated fatty acids keeps them supple at sea-water temperatures because unsaturated fatty acids have lower melting points than saturated ones. The general occurrence of unsaturated fatty acids among parasites of mammalian hosts cannot serve the same purpose because of their higher body temperature. However, the presence of such fatty acids in immature or larval forms that often have to invade cold-blooded organisms may be a temperature adaptation.

The remaining two groups of lipids, the phospholipids and the unsaponifiable sterols and terpenes have also been most intensively studied in the filarial worms. In *Dirofilaria immitis* the major lipid classes present are the sterols and their esters. Total phospholipid makes up about 0.7% of the wet weight of the worm and, of this, phosphatidylcholine and phosphatidylethanolamine comprise 87%. Qualitative analysis of the lipids in dog plasma shows that there is little correlation with the filarial lipid pattern, which more nearly resembles those of other nematodes.

Glycerol uptake by *D. immitis* is slow and plays little part in energy metabolism. Glycerol is incorporated into phosphoglycerides, diglycerides and free cholesterol. Acetate and oleic acid are more readily used than glycerol and, when radiocarbon-labelled substrates are used, radioactivity is found in phospholipids, cholesterol and its esters, di- and triglycerides and sterol esters.

A number of phospholipids have been detected in *D. immitis,* and their syntheses have been examined in some detail. Phosphatidylserine synthesis occurs in extracts from adult females but only by way of an exchange reaction, catalysed by a Ca^{2+}-dependent enzyme, in which L-serine replaces one of the bases in preformed phospholipid, phosphatidylcholine for example:

$$phosphatidylcholine + serine \rightarrow phosphatidylserine + choline$$

The rate of incorporation is low compared with that in the rat, which has an alternative synthetic pathway.

Phosphoethanolamine synthesis also takes place by an exchange reaction, in which ethanolamine replaces choline or serine in phosphatidylcholine or phosphatidylserine, but this is very slow, about 0.3% of the rate by which phosphoethanolamine is formed by the simple decarboxylation of phosphatidylserine. Decarboxylation is also about 30 times faster than a third synthetic reaction sequence which involves the Mg^{2+}-dependent enzyme, ethanolamine phosphotransferase. It catalyses the following sequence of reactions:

1. ethanolamine + ATP $\rightarrow$ phosphoethanolamine + ADP

2. phosphoethanolamine + CTP $\rightarrow$ CDP-ethanolamine + PP_i

3. CDP-ethanolamine + 1,2-diacylglycerol $\rightarrow$ phosphatidylethanolamine + CMP

(ATP, ADP – adenosine tri- and diphosphate; CTP, CDP, CMP – cytidine tri-, di- and monophosphate; PP_i – inorganic pyrophosphate).

Phosphatidylcholine can be synthesised by an analogous pathway and also by a S-adenosylmethionine-mediated methylation of phosphatidylethanolamine. Phosphatidylinositol can be synthesised by base exchange or by a *de novo* synthetic pathway. There is also a mechanism for synthesising phosphatidylglycerol, but no evidence for cardiolipin synthesis.

The synthesis of isoprenoid derivatives by filarial worms has recently received intensive study. There are many compounds in living organisms whose structures are apparently based on isoprene (2-methyl butadiene):

$$\begin{array}{c} CH_3 \\ | \\ CH_2{=}C{-}CH{=}CH_2 \end{array}$$

These compounds contain multiples of five carbon atoms so arranged that they can be notionally separated into isoprene-like fragments. They are called terpenes, and are found in oils like camphor and geraniol, in rubber, in carotenes from plants. They are also found as side chains in various vitamins and cofactors for biochemical reactions, such as the family of ubiquinones that are important in the respiratory electron transport system in mitochondria and in folate metabolism. A ubiquinone has the structure:

$$\begin{array}{c} CH_3 \\ | \\ \text{quinone} - (CH_2{-}CH{=}C{-}CH_2)_n\,H \end{array}$$

where n refers to the number of isoprene units in the side chain. Ubiquinones of different lengths are referred to as ubiquinone-9 or -10 and so on, depending on the number of units that they possess. There are many compounds with isoprenoid side chains present in nematodes. They include ubiquinone itself, cholesterol, rhodoquinone, farnesol-like compounds and the polypropanol, solanesol.

One important metabolic pathway leads to the synthesis of sterols. In adult filariae *in vitro* the starting point is mevalonic acid. They are unable to carry out the two terminal reactions to synthesise squalene or sterols. Neither can they add the isoprenoid tail to menadione in the synthesis of vitamin K_2 from menadione. However, they are able to synthesise ubiquinone-9, a number of dolichol isoprenologues and, predominantly, a short chain isoprenoid alcohol, geranyl geraniol.

These findings are not consistent with earlier observations that a small amount of cholesterol was formed after incubation with labelled glycerol or acetate, but they are more consistent with a number of other observations. For example, there was a barely detectable incorporation of radiocarbon from mevalonate, the preferred substrate, into squalene and cholesterol. It is therefore probable that *D. immitis* does not carry out the *de novo* synthesis of sterols from acetate. This view is supported by the much greater incorporation into isoprenoids, which is generally taken to imply the inability to make sterols.

Members of an interesting group of compounds, the ecdysteroids, have recently been detected in *D. immitis* and other helminths (see Mercer 1985). In insects, moulting and metamorphosis is controlled by ecdysteroid hormones and, of course, nematodes also undergo several moults during their development. It is not clear whether ecdysone is involved in nematode moulting, nor even whether the Class Insecta can be used as a good analogy for nematode endocrinology. Both male and female *D. immitis* contain ecdysone, 20-hydroxyecdysone, 20,26-dihydroxyecdysone and possibly ponasterone A. Generally, the concentrations are lower than those normally encountered in insects.

The discovery of dolichols in *D. immitis* is not surprising, as they are very widely distributed in invertebrates. Dolichols are implicated as cofactors in the transfer of sugars to glycoproteins and proteoglycans and, as microfilariae have a surface coat that is mainly glycoprotein, may be of great importance in the synthesis of cuticular components. It is therefore significant that the isolation of a glycosyl transferase in which dolichol phosphate promoted the transfer of glycosyl residues from sugar nucleotides to form glycoproteins has been reported (Comley, Jaffe and Chrin 1982). This enzyme is located in the microsome fraction derived from homogenates of *D. immitis*, and was able to transfer mannose from guanosine diphosphomannose to dolichol monophosphate in the presence of a divalent cation. The microsomes also promote the synthesis of a glycoprotein from mannose and α-lactalbumin.

3.4 Purines, pyrimidines and their salvage

Purines and pyrimidines are heterocyclic bases. They are universally distributed in nature. Adenine and guanine are derivatives of purine and thymine, cytosine and uracil are derivatives of pyrimidine. They commonly occur as nucleosides, in combination with a sugar, often D-ribose, or as nucleotides, in combination with a sugar phosphate. Adenosine triphosphate is probably the best known nucleotide. They contribute intimately to the structure of the deoxyribose- and ribose-nucleic acids and provide the basis for the genetic code. When they are incorporated into coenzymes and vitamins, they participate in numerous metabolic reactions. In the form of cyclic nucleotides, they act as specific signallers within the cell.

There is almost no evidence to suggest that either the parasitic Protozoa or the parasitic helminths are capable of synthesising the purine ring out of simple precursors. As Figure 3.5 shows, the precursors of purine synthesis in organisms that do have the capacity are carbon dioxide, formate, glycine, aspartate and glutamine. The accepted way of demonstrating synthesis is to include a carbon-labelled precursor into an incubation medium containing the parasite and then to analyse extracted purines for the presence of the label. Only in one case, that of *Crithidia oncopelti* has some evidence been obtained for the *de novo* synthesis of purines in parasitic groups.

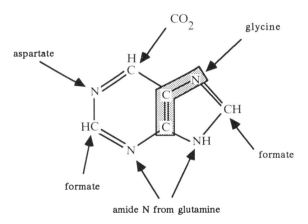

Figure 3.5: The Origins of the Atoms in the Purine Ring in *de novo* Synthesis

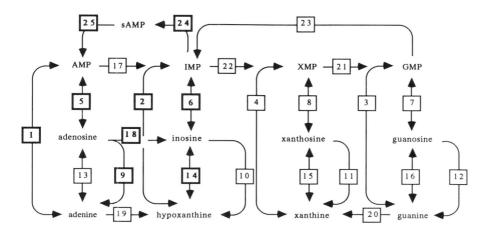

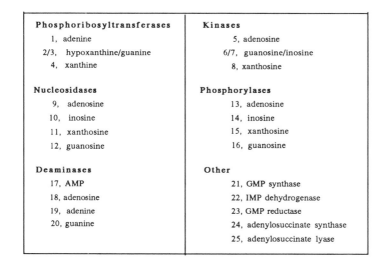

- enzymes with numbers in bold type have been detected in parasitic helminths. Different combinations of enzymes 1 - 25 have been detected in parasitic Protozoa but not all enzymes have been detected in a single organism.

AMP, GMP, IMP, XMP - adenosine, guanosine, inosine and xanthosine monophosphates, sAMP - succinyl AMP

Figure 3.6: Purine Salvage Pathways in Parasites

In the absence of *de novo* synthesis, it is not very surprising to discover that parasites have elaborate systems for securing and interconverting purine derivatives of host origin. Mechanisms for uptake have already been discussed (this Chapter). The reactions that result in the interconversion of purine derivatives are collectively known as salvage pathways.

Analyses of extracts of parasites show that they contain the same range of purines and purine derivatives that are found in free-living organisms. Experiments with parasite cultures in minimal media have, in many cases, demonstrated exactly which purines and purine-containing compounds are essential for growth. When the results of these experiments are combined with others in which radioactively labelled purines are employed, it is possible to assemble a fairly accurate picture of the salvage pathways in both parasitic protozoa and helminths. It is perhaps no coincidence that two groups of blood parasites are the best studied – this is no doubt because blood itself is a well studied and clearly defined medium. The two groups are the malaria parasites and the schistosomes. The pathways are illustrated in Figure 3.6.

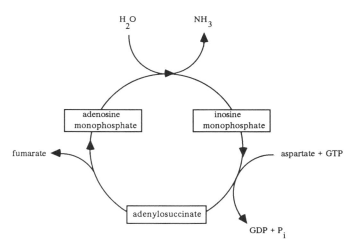

Figure 3.7: The Purine Nucleotide Cycle

One other pathway for the interconversion of purine nucleotides is known as the purine nucleotide cycle (Figure 3.7).

The overall effect of the cycle is summarised in the following equation:

$$aspartate + GTP + H_2O \rightarrow fumarate + NH_3 + GDP + P_i$$

which is an energy-dependent deamination of aspartate.

The enzymes for all these reactions are variously distributed in parasites, but while a clear role can be established for the salvage pathways in Figure 3.6, it is not known to what extent the purine nucleotide cycle is important. The latter is interesting because its clear association with respiratory metabolism (it utilises aspartate and GTP and generates fumarate) suggests that it could be controlled by GTP availability (generated, for example, by the PEPCK reaction) and competition for substrates by mitochondria.

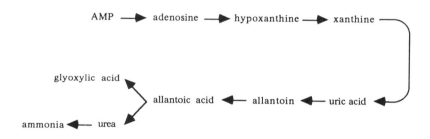

In this example, adenosine monophosphate (AMP) is the molecule degraded.

Figure 3.8: Pathway of Purine Degradation

Purines are not, of course, invariably salvaged. Nematodes possess a degradation pathway which can be inferred from the production of labelled intermediates and end products after the administration of labelled purine compounds to incubation media. It is a simple, linear pathway (Figure 3.8). There is no evidence for the presence of a urea cycle in parasitic helminths and it is likely that urea, sometimes detected in media in which worms have been incubated, is derived from the degradation pathway.

It is an attractive idea that the processes of purine salvage and the loss of *de novo* purine synthetic activity are parasitic adaptations that release the organism from the necessity of maintaining a large suite of enzymes to sustain a highly energy-demanding synthetic pathway. It is pure conjecture, however, and it is hard to see what sort of evidence might permit a clear resolution of the question. It seems very likely that the parasite groups once possessed these enzymes and subsequently lost them – indeed, comparative biochemistry of free-living relatives suggests that this is so. Biological systems are built on the principle of redundancy and it is difficult to understand the interplay of selection pressures that might lead to the total ablation of such a fundamental property of living matter as the ability to synthesise purines. But absence of evidence is not evidence of absence. It is possible that the genes that code for the enzymes of purine synthesis are present but not expressed. Much more work is needed to test this possibility.

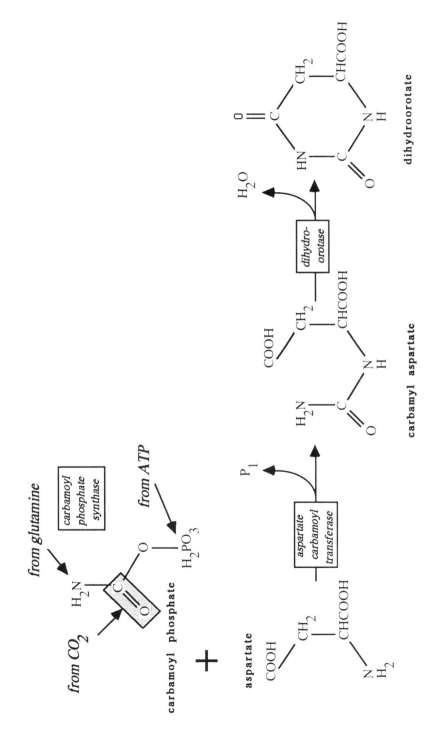

Figure 3.9: The *de novo* Synthesis of Pyrimidines

Here is a case where the parasite does indeed depend on its host for 'a minimum of one gene or its product' and yet it has to be said that such a clearly defined case is an exception. Why, for example, has not pyrimidine synthesis been subjected to the same selection pressures?

Most parasitic Protozoa and all helminths so far investigated seem to be capable of synthesising pyrimidines *de novo* from carbamoyl phosphate and aspartate. The sequence of reactions is illustrated in Figure 3.9.

The evidence for the pathway is largely circumstantial and derives from studies of parasite culture and the incorporation of labelled precursors into pyrimidines. Carbamoyl phosphate synthase is a key enzyme in pyrimidine synthesis and is now known to be present at low activities in many parasites where once it was thought to be absent.

There are also pyrimidine salvage pathways (Figure 3.10), but much less is known about them than those for purines. Their presence is inferred in parasitic helminths by the fact that pyrimidine bases are incorporated as nucleotides into nucleic acids.

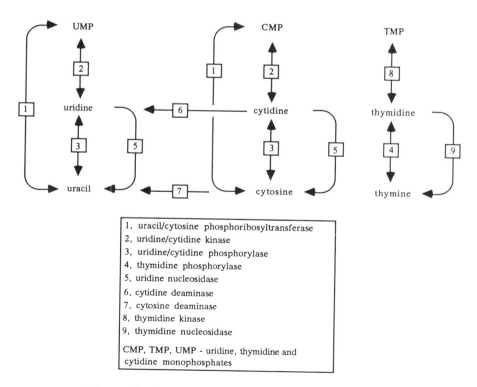

1, uracil/cytosine phosphoribosyltransferase
2, uridine/cytidine kinase
3, uridine/cytidine phosphorylase
4, thymidine phosphorylase
5, uridine nucleosidase
6, cytidine deaminase
7, cytosine deaminase
8, thymidine kinase
9, thymidine nucleosidase

CMP, TMP, UMP - uridine, thymidine and cytidine monophosphates

Figure 3.10: Pyrimidine Salvage Pathways

4
Host Immunity and Parasite Adaptation

4.1 Introduction

This chapter is concerned with the interaction between protozoan and helminth parasites and their hosts' immune systems. For the most part, we will emphasise parasites of vertebrates for which the term 'immunity' has special significance. It implies specific recognition of parasite antigens by particular types of host protein molecules: B-cell derived immunoglobulin and the somewhat more elusive T-cell receptor.

The relationships that exist between hosts and parasites are extraordinarily complex. Many of their details are difficult to comprehend and remain enigmatic. However, spectacular advances in a number of fields of endeavour – particularly recombinant DNA and hybridoma technologies, and immunological theory and techniques – have provided essential tools that make it possible to undertake the Herculean task of unravelling the elements of the parasitological puzzle. Until comparatively recently many immunological aspects of host-parasite relationships were consigned to the 'too hard' basket. But that attitude has quickly changed with the realisation that progress with even the most intractable problems can now be achieved.

This chapter deals with immunological aspects of host-parasite relationships at two levels; first, at a general, essentially evolutionary level; and then at a more specific level by considering some of the phenomena associated with selected examples of host-parasite relationships.

4.2 Evolution of multicellularity

Some writers (e.g. Hamilton 1982) believe that the evolution of multicellularity may have been, in part, an adaptive response to escape from parasites. Multicellularity certainly endows an organism with greater powers of dispersal and the opportunity to colonise new habitats. However, a multicellular body carries with it the penalty of a greatly increased generation time compared with potential pathogens, and it presents many new niches for exploitation by parasites. Even if the multicellular strategy were initially successful as a counter to parasitism it did not remain so as a vast and diverse assemblage of parasitic organisms testifies.

Multicellularity was clearly not an adequate strategy for avoiding parasitism.

Thus, hosts needed additional strategies to increase their genetic variation to match that generated by their rapidly reproducing parasites. In this way, some parity between their respective evolutionary rates could be obtained. Clearly, recombination of genetic material through sexual processes is one way of achieving this (Hamilton 1982), although we must not overlook the fact that most parasites also reproduce sexually at some stage of their life cycles. In addition, the evolution of multicellularity may have been accompanied by the acquisition of much larger genomes. Prospective hosts could thus draw upon greater reserves of genetic information in order to generate variation. It is also possible that greater opportunities for direct exchange of genetic material between hosts and parasites occurred (Howell 1985); this too could have had evolutionary implications which offset the differing generation times between the organisms involved.

4.3 Evolution of immune systems

One of the consequences that follows from the evolution of a multicellular body is the need for division of labour among its constituent cells. Following from that, however, is the fact that an animal is faced with the problem of defending many different types of cell. Moreover, the fundamental ability to distinguish self from non-self becomes a more demanding task because organisational stability has to be maintained among structurally and functionally diverse cells.

The basic ability of self/non-self recognition was almost certainly inherited by metazoans from their protozoan ancestors (exemplified among extant Protozoa in the form of, for example, conjugation in *Paramecium* and phagocytosis in *Amoeba*) and indeed was probably a precondition for the evolution of multicellularity. However, cellular specialisation and the greater degree of elaboration of body form and function provided the selection pressure for the emergence of a motile cell population, with the function of surveillance and attack of non-self invaders and of mutant cells of the organism that might threaten its integrity.

Throughout the evolution of the Metazoa the ancestral role of phagocytosis mediated by specialised phagocytic cells (variously referred to as amoebocytes or haemocytes in invertebrates and macrophages in vertebrates) has been retained as a principal defence mechanism. However, among vertebrates, a new class of cells, lymphocytes, appeared which endowed them with a more flexible and effective defence capability. This reaches its peak among homeothermic vertebrates (birds and mammals). The functions of lymphocytes form the basis of adaptive immunity. The advantages of adaptive immunity include:

(i) the provision of long-term memory of previous contact with foreign material;

(ii) the production of highly specific molecules (antibody or immunoglobulin produced by B-cells and the T-cell receptor which, according to Marrack and Kappler (1986), structurally resembles immunoglobulin) for recognition of foreign molecules (antigens or parts of them called epitopes);

(iii) the provision of a greater range of defences by interactions with macrophages;

(iv) the provision of protection for offspring during the time required for maturation of defence mechanisms in these young animals.

More detailed accounts of immunity and the complex cellular interactions involved are given by Klein (1982), Roitt (1984) and Marrack and Kappler (1986).

The evolution of defence mechanisms by multicellular hosts presumably resulted from the imposition of strong selection pressures by parasites. In turn, the defence mechanisms themselves must have imposed powerful selection pressures on parasites. Present day host-parasite relationships exhibit, from an immunological point of view, a diverse array of solutions to the problems posed for each of the participants.

4.4 Evolutionary outcomes of host/parasite immunological interaction

There are essentially three aspects to be considered in the immunological relationship between mammals and their parasites. These are: how the parasite presents itself to its host's immune system; how the host responds to this exposure, and how the parasite responds to the mechanisms the host brings to bear against it. Above all, whatever the mix in presentation and response the essential requirement of the parasite is the survival of a host population in which it can complete its life cycle and ensure transmission from one host to another. The host's major problem is to survive in the face of continued infection.

These considerations generate a number of evolutionary implications. First, the parasite may come to occupy habitats in or on the host where its antigens have greater or lesser exposure to the host's immune system. Second, the antigens of parasites may achieve varying degrees of complexity in number, origin and type. These antigens may be 'seen' in a multiplicity of ways by, and provoke widely differing responses from, the host's immune system. Third, parasites may evade these responses by a variety of strategies.

It is of course possible that, during the course of evolution, suitable accommodations may not have been achieved by some hosts and their parasites and their mutual extinction followed. Extinction of the host inevitably means extinction of its parasites, unless the parasite finds other suitable hosts. But the converse is not necessarily true – a host can, in theory, survive without its parasites, although some recent views of host-parasite co-evolution (Freeland 1983) claim that parasitised hosts may have an advantage in their interactions with nonparasitised competitors under certain conditions. The absence of parasites from a host species – that is, host insusceptibility to infection – is a phenomenon commonly termed 'natural' or 'innate' resistance. It may be sustained by many underlying processes – for example, particular aspects of host physiology or factors associated with transmission of the parasite. Innate resistance will not be considered any further here, but interested readers are referred to Mims (1982) and Wakelin (1984) for details.

Despite the huge variety of hosts and parasites, there is a relatively small number of possible evolutionary outcomes for the immunological relationships that develop between them. Each represents an attempt to solve the basic problem of such relationships – that of striking a balance between efforts by the host to limit infection and efforts by the parasite to counteract this control and to secure transmission to another host. The various possibilities are described briefly below. The first four of these have also been discussed by Cohen (1976) and Mitchell (1982a,b,c).

4.4.1 Concomitant immunity

Concomitant immunity occurs when a parasite infects a host and persists for a significant period, during which the host develops immunity to further infection. This apparent paradox is also called premunition. Clearly the host has evolved mechanisms which defend it successfully against secondary infections but which have no effect on the established parasites of the primary infection. There are advantages to both participants in this relationship: the host does not become overloaded with parasites (reducing intraspecific competition between the parasites also) while at the same time the primary parasites persist and continue to reproduce. Further benefits accrue to the parasite if the immunity generated is also effective against other species of parasites; interspecific competition is reduced.

There are several examples of concomitant immunity. Perhaps the most well-known are those of schistosome infections of mammals. There are two sources of evidence; that derived from laboratory infections and that from epidemiological data on human infections (Smithers and Doenkoff 1982; Butterworth et al. 1982; Capron et al. 1982).

Concomitant immunity is also seen in cestode infections of sheep and mice and in Fasciola hepatica infections in rats (Rickard and Howell 1982). Rodents recover from infection with protozoans such as Trypanosoma musculi and Plasmodium yoelii and are immune to reinfection. Since, in these infections, low numbers of parasites persist in locations such as the spleen and kidney, concomitant immunity clearly occurs (Zuckerman 1970; Viens et al. 1975).

4.4.2 *Sterilising immunity*

In sterilising immunity the host eliminates its parasite burden entirely and remains immune to further infection. The continued survival of the parasite population in such circumstances requires that it effects its transmission before rejection. Further the host population must be sufficiently large to permit the recruitment of susceptible individuals at the same rate at which others become immune. The rate at which individuals in the population become immune is less important if the parasites have long-lived free-living stages, or (i) dormant stages that survive and remain infective in the external environment, or (ii) intermediate hosts within which they survive for protracted periods.

Sterilising immunity has long been recognised in bacterial and viral infections. It has not been so clearly identified as a result of protozoan and helminth infections but there are a few notable examples. Perhaps the classic case is that of *Nippostrongylus brasiliensis* in rodents (Kassai 1982) and there are some recorded instances among *Plasmodium* species (Zuckerman 1963) and trypanosomes in fish (Woo 1981).

There are suggestions that sterile immunity follows recovery from infection with some *Plasmodium* species in rodents. However, following splenectomy of the host the infections may relapse, which indicates that concomitant immunity rather than sterile immunity operates in these cases (Zuckerman 1970).

4.4.3 *Modulating immunity*

Modulating immunity is a kind of partial immunity. It occurs when the host modifies the parasite's activities. There may be some effect on the parasite's rate of growth, egg production, migratory activity, tissue localisation, antigenic properties or constitution during the course of primary or challenge infections. In contrast to concomitant immunity and sterilising immunity absolute resistance is not achieved. It is as if the parasite were held in check. Modulating immunity could therefore represent that balanced state that many parasitologists perceive as 'ideal'. Perhaps the most well-known examples of modulating immunity are seen among trypanosome infections of mammals. In *Trypanosoma lewisi* in rats, parasite reproduction is inhibited by an antibody called ablastin but the parasites persist for some time in the blood before they are cleared (Taliaferro 1932). In *T. brucei* infections, parasites are periodically cleared from the blood by antibody but this is followed by the re-emergence of antigenic variants of the parasite (Steinart and Pays 1986). Each variant is unique to the host's experience and requires a specific antibody response to remove it. The net effect is an oscillating parasitaemia over a prolonged period.

Of course, host influences on parasite activities may sometimes be due to non-immunological factors (for example hormonal changes). The periparturient rise in faecal egg counts of sheep infected with nematodes is an excellent case. Moreover, it may not be possible at times to discrimate between effects on the parasite's activities brought about by either competitive interactions between members of the parasite population or the modulating influence of the host's immune response. Thus, unequivocal demonstration of modulating immunity will always be a difficult task, much more demanding than the demonstration of sterilising or concomitant immunity.

Clearly, modulating immunity presents obvious advantages to both host and parasite – the host is not overwhelmed by infection and the parasite's survival is thus not prejudiced. In addition, the opportunities for the parasite to achieve transmission may be increased.

While the categories of immunity described above confer perceptible advantages on both the host and parasite, the evolutionary paths which led to them are by no means obvious. For these types of immunity to have arisen at all, the fitness of at least some of the infected ancestors of modern hosts must not have been impaired by comparison with that of uninfected members of the same population. If it were, naturally resistant individuals would tend to spread in the population. Space limitations preclude any detailed discussion of the population genetics of host/parasite interactions which lead to the acquisition of immunity, and indeed the information on this topic is of bewildering complexity. However, some of the factors which may have been important are: the genetic heterogeneity of host and parasite populations; the proportion of hosts infected; the range of susceptibilities to infection among infected hosts; the relative degrees of immunological responsiveness among hosts; the host's past evolutionary experience of the particular parasite as well as other pathogens; the range of susceptibilities among parasites to host responses; the virulence of the parasite; the parasite's predilection site; and the degree of complexity of its life cycle (migratory activities in host tissues, number of intermediate hosts, whether it has free-living stages and so on). It also should be borne in mind that the total parasite population is a collection of sub-populations. The nature of, and the interactions between, the genotypes of the sub-populations add a further degree of complexity to the evolutionary process. Further discussion of host/parasite coevolution with respect to some of these factors is provided by May and Anderson (1983), Levin *et al.* (1982) and Wakelin (1984).

4.4.4 *Ineffective immunity*

In this case the host is chronically infected but no discernible immunity (in the sense of Sections 4.4.1–4.4.3 above) develops and infections become superimposed on existing ones. Essentially the host remains fully susceptible to repeated infections. A seemingly clear example in this category is that of *Fasciola hepatica* infection in sheep (Rickard and Howell 1982), where the host remains fully susceptible even though it responds in an immunologically specific way towards the parasite. This example may not be particularly good for supporting the notion that ineffective immunity could evolve among natural populations of hosts and parasites. Selective breeding of sheep may have had the side effect of eliminating any immunity that the ancestors of domesticated sheep exhibited towards *F. hepatica*.

It should be noted that the absence of any effective control over parasite activity offers the prospect of overwhelming infections developing (either in terms of parasite numbers or parasite virulence) with the result that the host population will be eliminated. However, the maintenance both of highly susceptible hosts and of prolific and virulent parasites is an evolutionary possibility (May and Anderson 1983). It is largely sustainable by the interplay between the transmission efficiency and the reproductive success of the parasite and the highly polymorphic nature of the genetic system that controls the host's response to infection. The notion of regarding a parasite as well adapted *only* if it does little harm to the host – an oft-quoted maxim in parasitology – is gradually losing favour. Provided the parasite's *transmission* is unimpaired by its virulence (i.e. its fitness in the long term is not lowered) the degree of harm or not becomes a largely irrelevant consideration. A further factor to bear in mind when considering why *F. hepatica* is so damaging to sheep is that it infects a large number of hosts whose susceptibilities vary over a wide spectrum (Boray 1967). This adds enormously to the inherent genetic heterogeneity in the system; high susceptibility in one host may be matched by low susceptibility in another.

4.4.5 *Unresponsiveness*

In this case, the parasite infects the host but at levels that do not elicit an immune response. In a purely immunological sense, this defines the state of tolerance.

Tolerance can take two forms – 'low zone' and 'high zone' (Roitt 1984). It could thus arise in low and high levels of infection in a host, that is by low and high levels of antigenic stimulation, respectively. The latter could be considered maladaptive in that it might lead to death of the host through parasite overload; but there seems to be no *a priori* reason why the parasite's life cycle strategies could not be geared towards very low, but chronic, levels of infection that do not provoke the immune system nor have any marked effects on host fitness. There is no recorded instance of tolerance of this kind in any parasitic infection (protozoan or helminth) in adult animals. However, infection during foetal life with *Toxoplasma gondii* and *Trypanosoma cruzi* may induce tolerance (Cohen 1976), and similar phenomena may lead to tolerance of filarial infections (Selkirk *et al.* 1986). In these cases, hyporesponsiveness may be beneficial to the host if the pathological consequences of infection are exacerbated by underlying immunological factors.

4.5 Population characteristics of parasitic infections

No host/parasite relationship necessarily conforms rigidly to any single category of immunity. There are two main reasons for this. First, infected hosts may oscillate from one category to another under the influence of extrinsic and intrinsic factors. For example, the periparturient rise in nematode egg counts from infected ewes is associated with hormonal and immunological changes during pregnancy and parturition. Thus, the host response changes from a modulating one to that of a less effective, possibly ineffective, response. Again, the host may be chronically infected at a low level, a state of affairs sustained by either concomitant immunity or modulating immunity. A temporary slackening of resistance by the host or an antigenic change in the parasite may result in an increase in the parasite's population density. Good examples are a malarial relapse or recrudescence, or the emergence of a new antigenic variant of a trypanosome. The point to emphasise, then, is that a host's immunological status may vary throughout its life.

Second, both host and parasite populations exhibit a great deal of genetic variability (Wakelin 1978a, 1984; Sher and Scott 1982; Mitchell *et al.* 1982; Anderson and May 1982, 1985a, 1985b). Thus, it seems highly likely that all the categories of immunity listed above could coexist among populations of hosts and their parasites. The relative mix of each could well be dependent on a wide range of selection pressures (immunological and non-immunological) that operate on both the host and the parasite, not only when they are associated during infection but at other times (for example when the parasite is present in its intermediate host or free-living). It seems that the genetic diversity of host and parasite populations militates against the host-parasite relationship establishing any particular type of permanent steady state and, indeed, tends to ensure that the range of possibilities outlined above is sustained.

One of the features of host-parasite relationships is that relatively few members of the host population harbour the major part of the total parasite population (Anderson 1978). This phenomenon is called 'overdispersion' and is illustrated in Figure 4.1; the relationship is close to a negative binomial distribution (Anderson and May 1978).

If it is assumed that the probability of transmission of a parasite to all individuals of the host population is the same, the above distribution is due to (i) differences in susceptibility and responsiveness among the hosts and/or to (ii) differences in infectivity and immunogenicity in the parasites. Thus, point A (uninfected hosts) represents hosts displaying innate resistance and those exhibiting sterile immunity together with the susceptible but (as yet) uninfected individuals present in the host population. At points B and C either concomitant immunity or modulating immunity is operative, holding parasite burdens in check. Point D represents that group of hosts displaying ineffective immunity (or high zone tolerance?) and consequently higher parasite burdens. Similar views on the heterogeneity of infection levels and responses within the host population have been expressed by Wakelin (1978b) and Mitchell *et al.* (1982). Wakelin (1984) has suggested that survival of parasites within the low responder cohort of the population could provide an effective immune evasion mechanism at the population level.

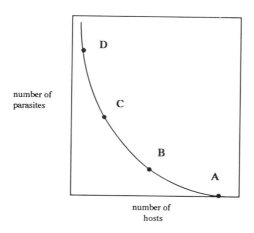

For an explanation of the points A, B, C, and D, see text.

Figure 4.1: Intensity of Parasite Infection among the Members of a Host Population

It is impossible to say to what extent the foregoing analysis of the immunological status of host populations is true. There are as yet no methods whereby immune status can be reliably assessed among the individuals of free-ranging natural populations and it seems likely that further progress in this area will be slow to develop. However, some general inferences can be made about the immune status of human populations with respect to malarial infection, based on epidemiological parameters such as degree of endemicity, and on clinical signs such as splenomegaly and parasitaemia in particular age groups in the population. For example, in areas of high endemicity adult individuals tend to exhibit significant degrees of protective immunity (Brown 1969). This is either concomitant or modulating immunity rather than sterile immunity because parasites are occasionally seen in blood smears. There are also some immunological tests that can identify humans possessing varying degrees of protective immunity to malaria (Perlmann 1986). Moreover, innate resistance to malaria is frequently associated with red cell characteristics such as the absence of Duffy blood group antigens, glucose-6-phosphate dehydrogenase deficiency, sickling and β-thalassaemia (Friedman and Trager 1981), and certain MHC antigens are associated with an ability to mount higher antibody responses in immune (or at least partially immune) individuals (Osoba *et al.* 1979).

As a further example it is possible to identify groups of laboratory-raised sheep with either high or low responsiveness to trichostrongyle nematodes as determined by the number of worms that establish in the gut of vaccinated animals (Dineen *et al.* 1978). At present this requires prior vaccination of the sheep with irradiated larvae followed by experimental challenge of the animal with normal larvae. The likely immune status of naturally infected or uninfected sheep cannot be gauged by other criteria at present. The ability to tissue-type these animals may assist in the future (Outteridge *et al.* 1984) and haemoglobin typing enables predictions to be made about their parasite burdens (Dineen 1984).

Other correlates of resistance to parasitic infection exist (Geczy 1984). For example, the HLA-B major histocompatibility locus is more strongly associated with high endemicity of malaria (and thus the presence of clinical immunity) than the HLA-A locus. Further progress in this area may lead to finer levels of discrimination.

4.6 Immunological interaction between hosts and parasites

It may be difficult to determine which of the categories of immunity referred to above operates in a given host-parasite relationship. Thus, we will now consider examples of host/parasite immunological interactions independently of these categories, but will refer to them where appropriate. In this section, we will focus on fundamental elements of the host response and the adaptive strategies the parasite employs to cope with them. Much of the information has come from *in vitro* and *in vivo* studies using clearly defined strains of laboratory animals, particularly rodents, as hosts.

In vitro culture methods are now sufficiently well developed with some parasites that extensive and sophisticated investigations can be carried out (Taylor and Baker 1987). Study of laboratory models of parasitic infections can suggest hypotheses about host-parasite interactions in natural populations, but extrapolating results to the field is another matter and testing hypotheses is extremely difficult. We need progress in this area to fulfil the objectives of controlling at least some parasitic diseases by vaccination. Progress will, however, tax the ingenuity of parasitologists for decades to come. Space limitations preclude a comprehensive coverage of the literature on immunoparasitology. More detailed information can be obtained from Cohen and Sadun (1976), Mitchell (1979 a,b,c), Barriga (1981), Mansfield (1981), Cohen and Warren (1982), Frank (1982), Owen (1982), Mitchell and Anders (1981), Mitchell *et al.* (1982), Wakelin (1984) and Parkhouse (1985).

4.6.1 *Protozoa*

Macrophage inhabitants

The macrophage is a most important and aggressive cell involved in both the afferent and efferent limbs of the immune response. Its choice as a habitat by parasites raises interesting and complex problems. Three examples are considered briefly here; more detailed accounts are given by Thorne and Blackwell (1983) and Mauel (1984).

(a) *Toxoplasma gondii* . The biology and immunology of this species has been reviewed in detail recently by Krahenbuhl and Remington (1982). *T. gondii* is a very widespread parasite which occurs in three forms during its life cycle – the tachyzoite and tissue cyst in the intermediate host and the oocyst in the cat. In its tachyzoite phase, it can invade almost any mammalian cell type; entry into macrophages is achieved by phagocytosis, which is presumably initiated by some form of binding between receptors on the cell surface and ligands in the parasite plasma membrane or glycocalyx. How other cells are infected is not clearly established (see discussion below on *Leishmania* spp). The host response (antibody and cellular components) is partially effective, in that all parasites other than those in intracellular sites in tissue (which are almost exclusively tissue cysts) are killed. Resistance to reinfection is established; the relationship between host and parasite is one of concomitant immunity since tissue cysts (and perhaps tachyzoites), derived from the primary infection, persist. Interestingly, this resistance is non-specific, as it will confer protection against infection with phylogenetically unrelated organisms, and this may have adaptive value in that the total host parasite burden is minimised. A principal effector mechanism in resistance to *T. gondii* appears to be the macrophage (perhaps natural killer cells as well), whose properties are radically altered on activation (Sethi 1982; David 1984); elevated levels of hydrogen peroxide (up to 25 times those of normal macrophages) and superoxide are implicated in killing the parasite. These chemical species interact to produce singlet oxygen, and hydroxyl radicals which are highly toxic.

Questions raised by the parasite's biology include the following:

(i) How does the parasite survive in macrophages in the early stages of infection? First, normal macrophages produce only low levels of peroxide and free radicals and the oxidative burst which usually accompanies phagocytosis does not occur when *T. gondii* is ingested. The parasite contains high levels of catalase which would contribute to the destruction of hydrogen peroxide and prevent its interaction with superoxide. Second, the vacuole containing the phagocytosed parasite (the parasitophorous vacuole) does not fuse with lysosomes but the reasons for this are not established. They are believed to relate to biochemical properties of the parasite's surface since if this is coated with antibody the parasitophorous vacuole fuses with the lysosome and the parasite is killed.

(ii) How does the parasite survive in the immune host? It has been suggested that survival in cells (other than in activated macrophages which kill the parasite) and the persistence of tissue cysts in the brain especially reflects the immunologically privileged nature of these habitats. Indeed, there is evidence that potentially lethal antibody cannot reach intracellular parasites and the existence of the blood/brain barrier to immune effectors is well known. However, there is some doubt as to whether higher levels of infection persist in the brain compared with cardiac and skeletal muscle.

(b) *Leishmania* spp. There are numerous species (or rather species complexes) of *Leishmania* (Wakelin 1984; Chang and Bray 1985; Blackwell, McMahon-Pratt and Shaw 1986) which infect man and other animals. The parasites multiply within macrophages. Essentially, there are two forms of disease caused by these parasites: cutaneous leishmaniasis (*L. brasiliensis*, *L. tropica*, *L. major* and *L. mexicana* occur in man, and the last two are readily established in mice; *L. enrietti* occurs in guinea pigs); and visceral leishmaniasis (*L. donovani* in man and transmissible to mice and hamsters).

It is difficult to make any general statements about immunity to these parasites and the reader should refer to recent reviews (Mauel and Behin 1982; Wakelin 1984; Chang and Bray 1985) for further details of immunoglobulins produced and the pathology of infection. The brief outline below gives some indication of the complexity of host-parasite interactions, with particular emphasis on the macrophage.

Studies on *L. major* (Mitchell and Handman 1985) indicate that T-cells in infected mice can have resistance-promoting or disease-promoting effects; the relative balance of each type of T-cell is determined by host genotype (via the MHC-complex) and thus the pattern of disease in different strains of mice can be remarkably different. It is argued that a lipid-containing glycoconjugated antigen of the parasite plays a central role in the disease. When this antigen is anchored in the macrophage and oriented in relation to certain Class II MHC molecules, it is able to induce a set of T-cells, T_{MA}, which in turn activate the macrophages (possibly by way of γ-interferon). Reactive oxygen intermediates are then produced which have deleterious consequences for the parasite. The number of T_{MA} cells is apparently regulated by the number of suppressor T-cells that are induced by the delipidised glycoconjugated antigen attached to a receptor on the macrophage surface – the more suppressor cells, the less effective resistance (i.e. fewer activated macrophages) is seen and the host may die (Balb/C mouse); with fewer suppressor cells, the disease can resolve (CBA/H mouse). This model may well be applicable in infections with other *Leishmania* spp, but the picture is as yet incomplete.

Phagocytosis of parasites by macrophages is a complex event – it involves interaction between complement (especially the C3b component) and a surface glycoprotein of the parasite which in turn interacts with inactivated C3b and mannose-fucose receptors on the macrophage surface (Blackwell *et al*. 1986).

Thus, in this instance the parasite makes positive use of an element of the host's defensive mechanisms to secure entry into the macrophage.

As with *T. gondii*, a number of adaptive strategies have to be invoked for continued survival and replication of the parasite within macrophages. First, as noted above, host genotype may affect the extent of macrophage activation. Second, amastigotes may interfere with the oxidative burst of macrophages via their surface acid phosphatase activity, or via a lipase which acts on the macrophage membrane, or through more generalised interference with macrophage physiology (Blackwell *et al*. 1986). The cutaneous-infecting species may achieve respite from the leishmanicidal activities of macrophages because of the impaired activity of these cells at temperatures below 37°C in the skin (Scott 1985).

Unlike *T. gondii*, phagocytosed *Leishmania* fuse with lysosomal vacuoles to form phagolysosomes. Thus, the parasites would seem to be at special risk of destruction by digestive enzymes of the lysosomes. It has been suggested that the amastigote membranes possess a proton pump which actively translocates protons from the parasite's cytoplasm into the lysosomal vacuole. This establishes an electrochemical gradient which drives the uptake of glucose and amino acids by the parasite. The effect of this pump may also be to lower the pH of the lysosomal contents and impair the activity of lysosomal enzymes. An alternative view is that 'megasomes', lysosomal-like organelles found within amastigotes, may play a role in parasite survival by modulating the phagolysosomal environment. Their effect may be to produce ammonia which raises the pH of the lysosomal vacuole and inhibits lysosomal enzymes. However, megasomes are not present in all species. In addition *Leishmania* spp. contain superoxide dismutases which are able to generate reduced glutathione, a free radical scavenger. These may provide an important defence against free radicals generated by the macrophage. Other mechanisms have also been advanced. There are so-called 'excreted factors' (negatively charged carbohydrates) that may have an inhibitory effect on lysosomal enzymes. Further, the surface properties of the amastigote may resist enzyme degradation, as the transformation from promastigote to amastigote following phagocytosis is accompanied by the acquisition of resistance to the digestive capacity of the macrophage. Immunosuppression may also occur and *L. donovani*, at least, can apparently seek refuge in fibroblasts.

(c) *Trypanosoma cruzi* *T. cruzi* gives rise to a chronic infection in humans. It can also infect many laboratory animals, the disease following a very similar course to that observed in humans. The parasite survives initial rounds of multiplication in macrophages and local tissue cells and circulates in the peripheral blood as a trypomastigote (acute phase). Subsequently, however, the parasite persists at low levels in the blood and tissues (chronic phase) and is clearly held in check by immunological mechanisms which also prevent reinfection. Thus, a state of concomitant immunity characterises the parasite's longer term relationship with its host. The immunology of the infection is comprehensively reviewed by Scott and Snary (1982). Factors that appear to be important in regulating infection are activated macrophages, antibodies and complement, and antibody-dependent cell mediated cytotoxicity (ADCC); cytotoxic T-cells and delayed hypersensitivity reactions are apparently of no major significance.

T. cruzi faces similar problems to other macrophage-inhabiting protozoans in avoiding destruction by these aggressive cells. In contrast to the species referred to above, phagocytosed *T. cruzi* trypomastigotes transform to amastigotes and escape from the phagosome into the surrounding cytoplasm where they divide freely. This location puts them out of the reach of lysosomes and lysosomal enzymes. Some protection within the intracellular habitat may also be derived from the parasite's ability to impair cell-mediated immunity, especially cytotoxic T-cell activity, which would otherwise attack parasite antigens expressed on the cell surface in association with host MHC antigens.

The bloodstream trypomastigotes are at risk from attack by antibodies which can prepare them for complement lysis, ADCC and enhanced destruction by activated macrophages. There are several suggestions as to how these immune effector mechanisms are either avoided or their potential to kill the parasite is at least diminished. These include:

(i) Non-specific immunosuppression by parasite-derived substances.

(ii) The protective nature of the parasite's surface (surface molecules with anti-phagocytic and anti-complement, especially anti-C3, activities).

(iii) The ability of parasite enzymes to cleave immunoglobulin at the cell surface leaving only Fab (antigen binding) bearing fragments attached, separated from the Fc (complement activating) fragments; this diminishes the prospect of both antibody-complement-mediated lysis as well as ADCC.

(iv) Capping and dislodgment of antibody-antigen complexes which form on the cell surface.

(v) Antigen sharing by the parasite and its host.

There is some evidence to back up all these possibilities, but a good deal of further work is required to give them greater support.

Erythrocyte inhabitants

(a) *Plasmodium* spp. There are more than 100 species of *Plasmodium* which infect a wide range of vertebrates. In this section, consideration will only be given to mammalian species. There has been an enormous upsurge of interest in malarial parasite immunology and extremely comprehensive recent reviews are available (Cohen 1979; Cohen and Lambert 1982; Wakelin 1984; Newbold 1985; Howard 1986). Problems that will be briefly addressed here concern the parasite's ability to survive despite the development by the host of a degree of resistance to reinfection – resistance based in part on antibody, cellular mechanisms such as ADCC, activated macrophages, cytotoxic T-cells, or cytotoxic mediators produced in response to T-cell interaction with malarial antigen, or non-specific responses evoked by plasmodial products. The stage (whether sporozoite, merozoite or gametocyte), species and strain specificity of antimalarial immunity strongly implies the involvement of specific effector mechanisms but, of course, interactions between these and other non-specific effectors cannot be ruled out.

Entry of parasites into red cells has parallels with phagocytosis by macrophages. The merozoite orients itself with its anterior end attached to the red cell membrane; the membrane invaginates and the merozoites become internalised in a vacuole (Aikawa et al. 1978). The process of entry is complex – receptors on the red cell membrane (e.g. Duffy blood antigens and glycophorins) and parasite antigens secreted from the rhoptry (an organelle at the anterior end of the merozoite) appear to be intimately involved, but precise details of the interactions which induce the membrane to invaginate are as yet, unknown (Howard and Miller 1981; Aikawa 1983).

Malarial parasites survive for considerable periods in otherwise immune hosts. Cohen and Lambert (1982) and Cohen (1984) consider that the following mechanisms, each probably contributing to some extent, are conducive to parasite survival:

(i) Within liver cells the parasite provokes no cellular reaction, presumably because antigens are not expressed on the cell surface. However, infected red cells express parasite antigens, which probably earmark them for removal by the kinds of mechanisms listed above; but red cells only weakly express MHC determinants, which may explain the lack of T-cell cytoxocity towards infected red cells. Such a phenomenon could require interaction between cytotoxic T-cells and parasite and MHC antigens of the same type as expressed by the T-cell (the MHC restriction requirement) on the infected cell surface. The range of immune effectors the host has at its disposal against *Plasmodium* is thus reduced.

(ii) Antigenic variation is apparent in some infected animals as recurrent peaks of parasitaemia, each peak representing a parasite population that is distinct from the peak which preceded it. By using the antigenic variation strategy, the parasite can exploit the lag between initial antigenic stimulation and the time taken for an effective response to be mounted by the host. When antigenic variation is taken together with the antigenic specificity of the different developmental stages of the parasite in the mammalian host, an effective parasite survival strategy is evident. The basis for antigenic variation in malarial parasites is unknown but, clearly, it is important to understand it if vaccination against the parasite is to become a reality.

(iii) Malarial parasites exhibit a profound influence on the host immune system, as evidenced by changes in the host spleen, lymph nodes and lymphocyte populations in the blood. In a variety of host species, immunosuppression may occur – and it may be general or specific. In the latter case, if responses which are crucial for the establishment of malarial immunity are affected, parasite survival is enhanced.

Another consequence of malarial infection is hypergamma-globulinaemia. For example, up to a 7-fold increase in IgG synthesis may occur; but only 5% of this may have specificity for the parasite. In the light of this it has been proposed that the parasite produces a mitogen which causes polyclonal activation of these B-cells. This can be viewed as a diversionary tactic, since only B-cell clones with specificity for the parasite should be activated if the host's defensive capabilities are to be optimised.

More recent work has suggested additional strategies which may enable the parasite to avoid the consequences of immunity.

First, several malarial antigen genes have been cloned and sequenced (Kemp et al. 1986; McIntyre et al. 1986). Many of their expressed products are associated with the surface of different life cycle stages or infected red cells (Nussenzweig and Nussenzweig 1985; Anders 1986; Targett and Sinden 1985; McIntyre et al. 1986). A striking feature of these antigen molecules is that they contain short repeated sequences of amino acids (depending on the antigen, four to thirteen amino acids may be involved) which are immunogenic; cross reactions occur between repeat sequences in different antigens. The repeats may be irrelevant to the development of protective host responses but their immunodominance may impair the host's ability to respond to 'host protective' epitopes on the same or other molecules.

Second, parasite antigens (referred to as S-antigens) appear on the red cell membrane at the time of rupture and release of merozoites (Howard et al. 1986). These may divert the attention of immune effectors from the merozoites at that time. Moreover so-called 'knob' proteins (produced by the parasite) on the red blood cell membrane may lead to sequestration of infected erythrocytes in capillaries and prevent them from reaching the spleen, where they might otherwise be cleared.

(b) *Babesia* spp. Extensive reviews of the immunological aspects of these red-cell inhabiting protozoans are available (Mahoney 1977; Callow and Dalgleish 1982; Jack and Ward 1981). In some respects the parasites resemble *Plasmodium* spp., apart from the fact that they have no extra-erythrocytic stages. In many cases, high levels of sterile immunity may follow a single infection. Until sterile immunity becomes established, the parasites may display antigenic variation (basis unknown) which prolongs their residence time in the peripheral blood. Other factors that may contribute to the parasite's survival are its apparent ability to incorporate (or synthesise?) 'host-like' antigens, bring about immunosuppression and deplete complement levels, possibly by secreting a protease. A dependence on complement for red cell invasion is borne out by the observations that: (i) mice deficient in C3 and C5 complement components are resistant to *Babesia* infection and (ii) the C3b receptor on red cells is critically involved in red cell recognition (and, perhaps, penetration) by some species. Like *Leishmania*, the parasite appears to exploit a component of its host's defensive armoury to facilitate its proliferation.

Extracellular protozoans

(a) *African trypanosomes.* Of all the parasitic Protozoa, the African trypanosomes (*Trypanosoma brucei* species complex, *T. congolense* and *T. vivax*) have been most intensively studied from the point of view of their mode of immune evasion. Many reviews of this topic are available (Borst and Cross 1982; Bernards *et al.* 1984; Hajduk 1984; Myler *et al.* 1984; Parsons *et al.* 1984; Seed *et al.* 1984; Turner 1982, 1984; Donelson and Turner 1985; Richards 1984; Steinart and Pays 1986) and only essential details will be covered below. On a broader front, detailed immunological aspects of these infections have been reviewed recently by Vickerman and Barry (1982) and Askonas (1984).

The parasites occur widely in game animals where they are not unduly pathogenic, but in man and domestic stock they cause more serious complications. Untreated infections give rise to a variety of symptoms including fever, anaemia, lymphadenopathy and lethargy, and are often fatal. The course of infection in both humans and animals is characterised by periodic rises and falls in parasite numbers. Each successive rise is due to the emergence of a new antigenic variant in the trypanosome population which is eventually controlled by an antibody response that is specific for each variant trypanosome population which emerges. The significance of antigenic variation is that it enables parasites to remain in the blood for long periods, with a greater probability of transmission to the vector. The pattern of infection in most hosts is thus one of modulating immunity – but, in the longer term, the host often dies and the immune response could be viewed in an overall sense as an ineffective one. Where clinical recovery from infection occurs, sterile immunity may prevail.

The key element in antigenic variation is a glycoprotein (with a molecular weight of 65 kD) which forms a monomolecular layer of several million molecules over the parasite's surface. This protein consists of a C-terminal carbohydrate-containing region which anchors it in the cell membrane, and a polypeptide chain (a sequence of about 360 amino acids) whose variability is responsible for the parasite's antigenic diversity. Thus, these molecules are referred to as variable surface glycoproteins or VSGs. Antigenic variation in trypanosomes is thus seen as the sequential expression of differing VSGs, presenting clinically as recurring peaks of parasitaemia and governed by the sequential appearance of antibodies specific for each VSG as it arises. Most of the trypanosome population at each point shows a common VSG and therefore conforms to a particular variant antigenic type or VAT. Antibodies are thought to play a selective rather than inductive role in antigenic variation and there is some degree of predictability in the sequence of the VSGs. A trypanosome may contain between 300-1000 VSG genes, so the potential repertoire of VSGs and, hence, successive waves of parasitaemia, is extremely large. The genetic mechanisms regulating the expression of VSGs are covered thoroughly in the above reviews and will not be considered further here.

While antigenic variation provides an effective immune evasion strategy for the parasite, immunosuppression is frequently observed in infected animals and this may also contribute to the parasite's survival. Immunosuppression (possibly macrophage-mediated) may be associated with polyclonal activation of B-cells by a trypanosome mitogen and consequential clonal exhaustion of trypanosome-specific B-cells, loss of T-cell function, low levels of complement, increased catabolism of immunoglobulins, antigenic competition and defective antigen processing.

(b) *Stercorarian trypanosomes.* The pattern of infection of rodents with *T. musculi* and *T lewisi* stands in marked contrast to that of African trypanosomes. There is no apparent antigenic variation of the African type and the parasites are cleared from the blood, after a single peak of parasitaemia, within a few weeks of infection (D'Alesandro 1970; D'Alesandro and Clarkson 1980; Vargas *et al.* 1984). Both antibodies and cells (particularly macrophages) may play a role in inhibiting parasite reproduction and promoting clearance, respectively, and the parasites do not appear to display any well-defined immune evasion strategies. Following clearance from the blood the host is resistant to re-infection but the parasite may persist in sites such as the kidney or liver. Thus, a state of concomitant immunity exists but whether this can progress towards sterile immunity is uncertain. It can be assumed that there has been no profound selection pressure on these trypanosomes to prolong their sojourn in the peripheral blood and thus their accessibility to the vector; the closer association of the vectors (fleas) of these species with the host than the vectors (tsetse flies) of the African trypanosomes may have some bearing on this question.

4.6.2 *Helminths*

Schistosomes
The immunological relationships between schistosomes and their definitive hosts have been much studied (see reviews by Smithers 1976, 1986; David and Butterworth 1979; McLaren 1980; Butterworth *et al.* 1982; Capron *et al.* 1982; Smithers and Doenkoff 1982; Capron and Capron 1986;). In several hosts (for example monkeys, mice and probably humans) a state of concomitant immunity exists, but rats eliminate their infections and therefore display sterile immunity. The antigens that provoke immunity are produced by all stages of development within the host (schistosomula through to adult) and are associated with the tegument. Several polypeptides, including some which are strongly glycosylated, are involved; the most important appear to be antigens with molecular weights of 16, 28, 32, 38 and 40 kD (Simpson and Cioli 1987; Balloul *et al* 1987).

Since all stages of development provoke immunity but only schistosomula are apparently vulnerable to immune effectors, there are two major questions: in immune animals (1) how are schistosomula killed, and (2) how do adult worms survive?

The reviews cited above cover at length the kinds of mechanisms that have been shown to kill schistosomula *in vitro* and *in vivo*. The mechanisms involved in various hosts apparently differ and the reviews should be consulted for further information.

Nearly every type of immune effector has been implicated in the killing of schistosomula *in vitro* – lethal antibody and complement, antibody in conjunction with eosinophils (requiring mast cells as accessory cells), or normal macrophages, or platelets (ADCC mechanisms) and neutrophils with complement. In addition, activated macrophages and their products, and phytohaemagglutinin-stimulated T-cells will kill schistosomula. The role of these various effectors *in vivo* is not clear-cut. There is evidence for antibody and eosinophil involvement.

Schistosomula become resistant to the effects of immune effectors after a short period (1-4 days) of development in previously uninfected animals or after cultivation *in vitro* for 24-48 h. A number of immune evasion mechanisms are apparently deployed by the parasite during this period, and these persist throughout the parasite's subsequent development and residence in the host. By the time the host with a primary infection has mobilised its immune system, the parasite is securely protected.

A number of mechanisms that may be involved in the evasion of immunity have been postulated (Clegg 1972, 1974; McLaren 1984 and other references listed above). These include:

(i) The adsorption of host macromolecules (so-called host antigens, such as glycolipids, MHC antigens, red cell glycoproteins and immunoglobulins) which disguise the parasite as host tissue or reduce the level of exposure of its surface antigens. There is considerable debate as to whether the parasite has the ability to synthesise 'host' molecules but there has been no unequivocal resolution of the question; evidence favours adsorption from the host, but the synthesis of some 'host' molecules – mouse α_2-macroglobulin for example – whose function is obscure, cannot be ignored. It is possible that glycosylating enzymes of the parasite are involved in the acquisition of host antigens since one of the common features of these molecules on the tegument (with the exception of immunoglobulins) is the presence of carbohydrate moieties.

(ii) The rapid turnover rate of the tegument (2-4 h) and the development of its unique heptalaminate structure soon after infection. The acquisition of these properties coincides with the loss of susceptibility of schistosomula to immune effectors.

(iii) Proteolytic cleavage of immunoglobulins, so that their Fab regions become separated from their Fc portions, which remain attached to the parasite via Fc receptors on the parasite surface. Essentially, immunoglobulins may bind in such a way that their Fab regions, containing the antigen combining sites, are oriented away from the surface antigens of the parasite. Moreover, the Fab fragments released are further broken down into small peptides and these may inhibit macrophage activity.

(iv) The production of schistosome derived inhibitory factor and schistosome anticomplementary antigen by the parasite, together with immune complexes depress the immunological potential of the host. Recently, a low molecular weight schistosome-derived substance has been described which potentiates T and B lymphocyte proliferation (Aurriault *et al* . 1984). This factor and the suppressive factors noted above and in (iii) give the parasite the capacity to regulate immune reactivity and to determine more precisely the level of host response.

Immunological phenomena occur in response to the eggs of the parasite, which become lodged in ectopic sites (particularly the liver) instead of passing to the exterior from the gut lumen or bladder. Granulomata development around eggs is a T-cell-dependent form of delayed-type hypersensitivity (Warren 1978). Macrophages and eosinophils are involved in the pathological consequences producing a variety of substances including prostaglandins and superoxide (Chensue *et al.* 1983). If granulomata formation is seen as an attempt by the host to localise damage caused to tissue by substances secreted from the egg (hepatotoxins and soluble egg antigens) – that is, having a protective function – the protective effect to the host must be weighed against the deleterious effects of the consequent pathology. There is evidence that granulomata size decreases during infection, due to a modulating effect of suppressor T-cells, and that egg release from tissues into the gut or bladder is optimised by responses to egg antigens. Taking these factors into account, the host response can be viewed as having survival value to both the host and the parasite.

Larval taeniid cestodes
The majority of the hosts of larval taeniid cestodes develop an impressive level of resistance to reinfection (see Williams 1982 and Rickard and Williams 1982, for reviews of the parasite species and detailed analyses of their hosts' resistance characteristics and immune mechanisms). The tissue stage (cysticercus, strobilocercus or hydatid cyst), however, survives for prolonged periods in the face of this resistance. Thus, concomitant immunity characterises infections with these species. There is appreciable cross resistance between different species of larval cestodes, which clearly reduces interspecific competition between them (Schad 1966).

Antibodies clearly play an important role in preventing re-infection, although ADCC mechanisms involving eosinophils and neutrophils (Beardsell and Howell 1984) may also be involved.

In a challenge infection, the target of these immune effectors is apparently the oncosphere as it penetrates the gut wall or during the earliest stages of its reorganisation following penetration. However, in a primary infection, immune evasion strategies are presumed to be set in place by the parasite before effector mechanisms can be brought into play. This would appear to be borne out by studies with various inbred strains of mice (Mitchell *et al.* 1982) which indicate that there is a genetically determined rate of response to infection with *Taenia taeniaeformis*. Hosts that respond slowly (CBA/H mice) are more susceptible and thus develop more cysts than rapid responders (C57BL mice); in the former, the parasite is believed to elaborate appropriate evasion strategies before a potentially lethal host response has developed.

The mechanisms which are thought to enable the post-oncospheral stages of the parasite to avoid the effects of immunity can be summarised as follows:

(i) the initial microvillous layer around reorganising oncospheres and cysts may isolate the parasite from immune effectors. Later the cyst wall may form a truly functional barrier, particularly to cellular effectors, although immunoglobulins may appear in the cyst fluid;

(ii) the presence of immunoglobulins on the parasite may have a masking or blocking function;

(iii) the production of anticomplementary factors may deplete local complement levels in the vicinity of the parasite and reduce the probability of the lytic attack sequence of the complement cascade being triggered on the membranes of the parasite.

Nematodes

Immunity to nematodes has been reviewed in detail by Ogilvie and De Savigny (1982), Miller (1984, 1986) and Wakelin (1984, 1986).

(a) *Intestinal species.* Essentially, hosts of these parasites may be classified as responsive or non-responsive – a reflection on the host's genetic capacity to mount an effective immune response. Immunity, where it develops, can bring about the elimination of worms and can be expressed either against infective larvae of a challenge infection as soon as they reach the gut (so-called rapid expulsion), or against established worms before they reach patency, or against adult worms after patency has been reached. The response may, however, not necessarily bring about worm elimination but may cause stunting, lowered fecundity or inhibition of larvae in the mucosa.

In non-responsive hosts, nematode survival is prolonged; frequently these hosts are young animals. For many infections the surviving nematodes are adults but the host is resistant to incoming infective larvae. When the response to these larvae is strongly expressed, the adult worms may also be eliminated by attendant inflammatory changes in the well-known self-cure phenomenon. In summary, in some cases a state of sterile immunity may prevail (e.g. *Nippostrongylus brasiliensis* in rats); in others concomitant immunity is exhibited (*Haemonchus contortus* in some breeds of sheep), while in yet others the host may display ineffective immunity (*Ascaris lumbricoides* in man). It is, of course, quite remarkable that effective responses to nematodes can develop in a location such as the gut. The parasites have a tough cuticle and are not so intimately in contact with immune effector mechanisms as blood or tissue parasites.

Rapid expulsion is believed to be brought about by parasite specific IgA in intestinal mucus and by IgE-coated mast cells. It is apparently triggered by the relatively small number of larvae of the challenge infection that reach the mucosa, with inflammatory consequences for the resident adult worm population. There may be a parallel here with certain non-specific effects that can compromise the survival of nematode species that are not antigenically related to nematodes that triggered the original response (Dineen and Wagland 1982).

The expulsion of established worms appears to involve a wide variety of immune effectors. T-cells are crucial to the development of immunity, and worm elimination is associated with the recruitment of lymphocytes (some producing IgE, IgA and IgG) and other cells (eosinophils, basophils, mast cells and macrophages) involved in inflammatory responses in the *lamina propria*, together with pathological changes of the epithelium (e.g. hyperplasia of crypts, increased numbers of goblet cells and mast cells, villus atrophy).

There is a growing body of knowledge concerning nematode antigens (Maizels *et al.* 1982; Almond and Parkhouse 1985). It has been clearly shown that the nematode cuticle is antigenic; various antigens are also released from orifices of the worm and during moulting. Secretions from the stichosome of *Trichinella spiralis*, metabolites of 4th stage larvae of *Trichostrongylus colubriformis*, and antigens derived from moulting 3rd stage larvae of *H. contortus* have conferred some immunity on their respective hosts.

There is little information on how parasites in chronic infections evade immune responses by the host. It is possible that, by virtue of their habitat, they fail to stimulate protective responses or are remote from potential effectors. There are suggestions that immunosuppression (perhaps by the induction of suppressor T-cells) is involved; antiphlogistic and other immunomodulatory substances may be secreted by adult worms.

(b) *Systemic species.* Several intestinal nematode species have life cycle stages that also infect body tissues of intermediate and definitive hosts (e.g. *Toxocara canis* and *Trichinella spiralis*). Other species only occur in tissue sites and attain sexual maturity there (e.g. filarial parasites).

The immunology of tissue inhabiting nematodes has been reviewed by Piessins and McKenzie (1982), Ogilvie and De Savigny (1982) and Wakelin (1984) and many of the effector mechanisms involved in infections with gut-dwelling nematodes are also implicated in systemic infections. They include T-cell dependency, ADCC, eosinophils, neutrophils, mast cells and elevated IgE levels. Interestingly, macrophages are apparently not of any major significance.

A common feature of infections with systemic nematodes is their chronic nature. Thus, the parasites must put in place effective immune evasion strategies if some degree of protective immunity develops. Host genotype can greatly influence responsiveness to these parasites.

Surface or cuticular antigens of the nematodes appear to be primarily responsible for stimulating immune effectors. Thus, the cuticle is the potential target of immunity, and it may be protected from attack by mechanisms similar to those in schistosome infections. There is some variation between species as to which host molecules are absorbed by the parasite (McLaren 1984) and the functional significance of the phenomenon has yet to be established.

Trichinella spiralis adult worms in the gut survive for relatively short periods compared with larvae in muscle cells. Since migratory larvae are susceptible to immune attack it can be assumed that the intracellular habitat of encysted larvae is an immunologically privileged site. How naked larvae, such as those of *Toxocara canis*, avoid the effects of immunity is unknown. However, they are evidently held in check by modulating immunity, as evidenced by their resumption of activity when immunological resistance in these animals falls (e.g. during pregnancy).

4.7 Conclusions

Our understanding of immune effector mechanisms against protozoan and helminth parasites has greatly outstripped our understanding of immune evasion strategies employed by parasites; the conceptual framework is there, but the design of experiments to test ideas is a difficult and complex task that taxes the ingenuity of immunoparasitologists.

We are still some way from understanding the afferent aspects of immune stimulation by parasite antigens. Some questions that arise are: what are the structural characteristics of the antigens involved, how are they processed and how do they influence interactions between macrophages and lymphocytes? Do parasite antigens have epitopes which exert differential effects on T-helper and T-suppressor activities? Do parasites produce molecules analagous to lymphokines which influence the ways in which host lymphocytes are sensitised to antigen? Perhaps further work in this area will reveal intriguing adaptive strategies on the part of both host and parasite. The discovery of immunoregulatory substances in schistosomes is a start in this direction; the mitogens that some protozoans are believed to produce is also an area worthy of further study.

Beyond the laboratory, the wider implications of host/parasite immunological interactions at the population level are rapidly becoming appreciated. The ability to 'immunotype' individuals among naturally infected populations will almost certainly be of great significance, particularly if vaccination is to become a practical reality as a control procedure. This question has been discussed in more detail recently by Perlmann (1986) in relation to malaria.

Given the progress made in the last 10 years, a continually clearer and more comprehensive picture of both host and parasite adaptations to the constraints each imposes upon the other should emerge rapidly. It would be presumptuous, however, to make any predictions about what lies in store!

5

Biochemical Adaptation and the 'Magic Bullet'

5.1 Introduction

It is useful in a discussion of biochemical 'adaptation' in parasites to consider how our knowledge can be applied to the control and treatment of parasitic infections. In theory, it should be possible, after a thorough examination of parasite metabolism, to detect metabolic 'weak points' and exploit them by chemotherapeutic means. In practice, however, the identification of weak points, or molecular 'targets' for chemotherapy, in parasite metabolism – when they have been identified at all – has generally resulted from retrospective studies on the modes of action of successful antiparasitic compounds, or from comparisons with other groups of pests such as bacteria or insects. In this way, we have gathered much useful information about parasite metabolism, but only in very limited areas. By widening the scope of our general background in parasite metabolism, we should be able to identify new areas, with new potential targets, that are worth closer investigation.

In this chapter we discuss chemotherapy from the point of view of surveying both potential and identified molecular 'targets' ('receptors' or sites of action) in parasites. We will examine, in particular, potential targets that arise from specialised aspects of parasite metabolism and we will also discuss those aspects of host and parasite metabolism and physiology that are responsible for the selective elimination of the parasites. At the end of the chapter we consider how some of the specialised features of metabolism in parasites are related to the parasitic habit.

5.2 Chemotherapy

'Folk' or traditional remedies for human and animal illness have been in use with variable success since ancient times and some traditional potions are still used widely throughout the world today. Indeed, the scientific basis for the apparent efficacy of some of these remedies is currently being actively investigated in the search for new compounds to treat parasitic and other diseases. Examples of useful traditional treatments for parasitic diseases include cinchona bark (which contains quinine) for treatment of malaria, ipecacuanha (containing emetine) for amoebiasis, santonin (an anthelmintic) from the dried flowers of *Artemisia* spp., male fern (aspidium, an anthelmintic) extracted from the rhizomes of the fern *Dryopteris filix-mas* and ascaridole (an anthelmintic) extracted from the plant *Chenopodium ambrosoides* L. var. *anthelminticum*. Although these and other 'treatments' have been employed for centuries, they were at the time not necessarily considered to attack the *cause* of the illness (except perhaps in the case of infestation with worms, which are rather more obvious than Protozoa or other microorganisms), but were thought in some way to assist the body to heal itself and were in general directed at alleviating specific *symptoms* of illness.

Historically, the concept of using specific remedies to treat the cause of a specific, identified disease did not appear until the nineteenth century. Advances in medical research, led by Robert Koch, permitted the precise identification of, and therefore diagnosis of, specific disease entities. At the same time, it was recognised that many diseases are caused by specific infective agents. It then became possible to consider treatment not only to alleviate the unpleasant symptoms of illness, but also to treat the underlying cause of the problem. The science of rational chemotherapy began in earnest with the work of Paul Ehrlich at the turn of this century, who invented the term 'chemotherapy' and advanced the concept of the 'magic bullet', an agent of low molecular weight that selectively attacks the foreign invading organisms causing the disease, without causing damage to the host.

Ehrlich pioneered the use of certain dyes, known from microscopy to stain particular microorganisms, cells or cellular organelles selectively, in the treatment of infection. This work led to his most well-known discovery, the efficacy of arsenic-containing organic compounds in the treatment of syphilis. Some of these arsenical compounds were also effective against trypanosomiasis. Ehrlich introduced the concept of 'receptors' for drugs and showed, for example, that the chemoreceptors for the arsenical compounds in trypanosomes were -SH groups, which reacted to form As-S bonds that led to the death of the parasites (Ehrlich 1909). Receptors are specific to sites at which molecules such as drugs can bind and exert an effect on a cell or organism. The idea of receptors for drugs slowly became firmly established and it has gradually been shown in subsequent years – for some drugs at least – that these receptors correspond at the molecular level to precise sites on enzymes, coenzymes, macromolecules or cell membranes. Considerable pharmacological research today is devoted to characterisation of specific receptors and design of compounds to interact with them.

5.2.1 *Selectivity*

Central to the concept of the 'magic bullet' is the property of selective activity against the invading organisms. Ehrlich introduced the idea of the *chemotherapeutic index,* which is the ratio

$$\frac{\text{maximal dose tolerated by the host}}{\text{minimal curative dose}}$$

and is a general guide to the selectivity and usefulness of a compound. As a general rule, the chemotherapeutic index for a compound must exceed a value of 3-5 to be tolerable in practice. Selectivity of antiparasitic or antimicrobial compounds for their targets is brought about by a variety of factors, which can be summarised as follows:

(a) *Differential distribution:* The drug, though toxic to both host and parasite, is accumulated by the parasite only via, for example, specific uptake, ingestion or pinocytosis. Many anionic compounds are bound to albumin in blood; they do not enter the majority of the host's cells but, once taken up by a parasite, are cleaved to release the active agent. Alternatively, the drug receptors in the host, but not the parasite, are located in a protected environment (e.g. the brain) to which the drug has no access. Some orally administered compounds may affect gut-dwelling parasites but have no effect on the host because they are not absorbed into the host's body.

(b) *Differential detoxification metabolism:* The drug, though toxic to both host and parasite, is rapidly detoxified by the host, whereas the parasite is unable to do this. This may involve the presence of detoxifying enzymes such as cytochrome P-450 in the host but not in the parasite.

(c) *Differential activation metabolism:* The parasite, but not the host, possesses the metabolic machinery to metabolise the compound to toxic products. This occurs within the body of the parasite and the host is unaffected.

(d) *Unique 'receptors':* The parasite possesses binding sites or targets for the compound that are not present in the host. The host is therefore unaffected by the compound.

(e) *Differential binding to 'receptors':* Although host and parasite possess similar 'receptors' for the drug, the kinetics of binding are different, such that the compound is bound more readily by the parasite's 'receptors' than by the host's.

(f) *Differential reliance on the function associated with the 'receptors':* The host is less dependent than the parasite on the function affected by the drug, or may possess more effective bypass mechanisms; thus, although the host is affected by the drug, it is not lethal or immobilising as it is for the parasite. In this category it is also possible that the 'receptor' belongs to the host and that the parasite is more dependent on the function associated with it than is the host; this would apply particularly to intracellular parasites.

(g) *Activation of host defence mechanisms:* The drug has no effect on the parasite but causes the host to mount a more efficient immunological/immunochemical attack on the parasite.

These categories are probably not exhaustive, neither are they exclusive: an effective compound probably has several of these grounds for selectivity. In addition to investigating the specific receptors in invading organisms, modern pharmacologists also explore the pharmacokinetics of useful compounds and modify them to alter retention and distribution in the body of the host, thus balancing the selectivity in favour of the host. This is especially important in categories (a), (b) and possibly (g) above.

Most antiparasitic compounds in use today, and many experimental ones also, have been discovered not by rational design and synthesis, but by massive empirical screens of naturally-occurring compounds and their synthetic relatives. The starting point for such screens and syntheses has sometimes been a traditional remedy (as in the case of quinine and its derivatives), or the vital dyes (suramin is an analogue of trypan red, mepacrine is related to methylene blue), but often effective compounds have been discovered unpredictably from secretions by microorganisms (e.g. avermectins) or extracts of plants (e.g. artemisinine). This is an expensive method of searching and it is the dream of medicinal chemists to have sufficient information about the receptors for drugs in important parasites to design appropriate compounds rationally and systematically. We are not at present in a position to do this. In fact, basic information about potential drug receptors in parasites is quite rudimentary and we are even uncertain about the molecular mode of action of many important antiparasitic compounds commonly used today. It is important that this situation be redressed so that new classes of compounds can be developed to combat the increasing occurrence of resistance to current drugs.

5.2.2 *Potential drug targets*

Because of their specialised modes of life, parasites have modifications of metabolism, development and structure that present potential targets for chemotherapeutic attack because they are different from their hosts. Parasitic Protozoa are in an evolutionary sense further removed from their hosts than the helminths and, like the helminths, are a diverse group with different evolutionary origins. It is not therefore possible to generalise about potential drug targets, but we can outline some general strategies for searching for suitable drug targets.

Parasitic Protozoa multiply rapidly in the host, as do most other invading microorganisms. Many species also undergo some form of maturation or differentiation in the host during their life cycle. Therefore, metabolism associated with rapid growth and cell division, such as nucleic acid, protein and membrane synthesis, and processes controlling maturation or development, such as polyamine metabolism, are potentially very suitable targets in this group. Most helminths, on the other hand, generally invade the vertebrate host at the adult or late larval stage of their life cycle and undergo maturation, but not proliferation.

In adult helminths, proliferation takes place in the specialised reproductive tissues and inhibiting such proliferation would not (with some exceptions) necessarily affect the disease process because it does not eliminate the parasite, though it may interrupt the life cycle. Therefore, potential targets associated with rapid growth and cell division in helminths may be less useful therapeutically. Suitable targets in helminths are more likely to include processes such as muscular activity and neuromuscular coordination, nutrient uptake, some anabolic pathways, interparasite communication and essential energy metabolism. Because helminths are Metazoa, they are more closely related (in an evolutionary sense) to their hosts. Therefore, it is perhaps more likely that unique receptors will be found in Protozoa than in helminths. It could be argued, for the same reason, that chemotherapeutic indices are likely to be lower for anthelmintic compounds, because similar 'receptors' are probably present in the host.

The diversity of Protozoa and helminths precludes generalisations about potential chemotherapeutic targets. It also precludes generalised drug screening and necessitates separate screening methods for each type of important parasite. For the same reason, only limited use can be made of experimental models or *in vitro* screening, because it is the parasite-host *system* that is ultimately to be treated. Consequently, screening is inefficient and expensive.

5.3 Drug targets in parasites

Potential drug targets in parasites can be roughly categorised according to the type of metabolism or process to be attacked. Unfortunately this is complicated because many compounds in current use appear to affect a number of processes in the target species and it is difficult to determine whether there is a single primary target or 'receptor' (and, if so, what it is) or whether, indeed, the drug affects a multitude of processes. Good examples of this problem are given by suramin, and the antimalarial agent chloroquine, as will become evident below. In addition, many compounds attack different taxonomic groups of parasites and there is no necessary justification for assuming that they work in the same way against each group. Therefore, each case has to be considered on its merits. Because our background information on metabolism and development in parasites is very limited, this survey of targets represents only the tip of a very poorly described iceberg.

5.3.1 *Energy metabolism*

Many parasites rely almost exclusively on the reactions of glycolysis to supply their ATP requirements, as discussed previously in Chapter 2. Because the energetic yield from glycolysis is considerably lower than that from fully oxidative mitochondrial metabolism, parasites have a very high glycolytic flux rate and consume considerable quantities of substrate (usually glucose or glycogen). In addition, some protozoan groups (the trypanosomatids) do not store carbohydrate energy reserves and therefore rely on a rapid uptake of substrate from their environment.   Bloodstream *T. brucei,* for example, consume 85 nmol glucose/min/mg protein, which is equivalent to their dry weight in glucose every 90 min; this rate of glycolysis is 50 times that of the host (Fairlamb and Bowman 1980).   Inhibition of glycolysis would therefore be devastating for such a parasite.

The metabolic control mechanisms that permit such high rates of flux are likely to be different from analogous mechanisms in the host, and some enzymes will probably have different properties.   Indeed, it has been shown that hexokinase and phosphofructokinase in *T. brucei* are not sensitive to the usual feedback regulators for these enzymes found in mammals (Nwagwu and Opperdoes 1982; Cronin and Tipton 1985).   Minor differences in regulatory properties of the glycolytic kinases have also been reported for many species of helminths (see Barrett 1981).   Thus, different glycolytic 'receptors' or targets probably exist and should be suitable for attack in these parasites.

5.3.2 *Glycolytic enzymes*

Inhibition of any enzyme in the glycolytic pathway will be detrimental to a parasite's survival if the TCA cycle reactions and associated pathways are not present or functioning.   The greatest effect at the lowest doses of an inhibitor will be achieved if the target enzyme catalyses a *rate-limiting* or *rate-controlling* step in the pathway.   Rate-limiting enzymes in a pathway are usually those subject to tight regulatory control by allosteric and other means, and are considered to be very precisely adapted for optimum function of the tissues in which they occur.   Thus, as discussed above, in parasites relying on a high flux through the glycolytic pathway, the control mechanisms for the rate-limiting enzymes might be quite different from those in their hosts.   The rate-limiting enzymes, therefore, are good targets for chemotherapeutic attack.

A special case of a rate-controlling enzyme is one that controls the 'flux-generating' step of a pathway such as glycolysis (Newsholme 1978).   These enzymes may be switched 'on' or 'off' by covalent modification (phosphorylation and dephosphorylation), and they determine the maximum possible flux through the pathway – assuming no limitation of substrate – when they are switched 'on'. Such rate-controlling enzymes could be particularly valuable targets for attack.   By analogy with mammalian muscle or liver systems (see Hochachka and Somero 1984), the 'flux-generating' steps for glycolysis in parasites will be either glucose uptake from the environment (i.e. the glucose transport system) or, in parasites that store glycogen, the first step in glycogen degradation.   This is catalysed by glycogen phosphorylase, whose activity is controlled by a cascade of phosphorylation-dephosphorylation enzymes ultimately affected by environmental (i.e. extracellular) signals.   We have little information about such control processes in Protozoa; in helminths, the glycogen cascade system appears to be qualitatively similar to that of the host. We consider that these enzymes, or ones with similar control functions for an entire pathway, are prime targets for chemotherapy.

A number of chemotherapeutic agents have been shown to affect glycogen metabolism in helminths. In *S. mansoni*, niridazole inhibits glycogen phosphorylase phosphatase, the enzyme responsible for inactivating glycogen phosphorylase *a*. This disrupts the control of glycogenolysis, which increases in rate, causing the worms to deplete their carbohydrate reserves and eventually starve (Bueding 1970; Bueding and Fisher 1966). Most host tissues are unaffected by this compound, though depletion of glycogen occurred to a limited extent in muscle.

Another anthelmintic affecting the control of glycogen metabolism is levamisole, which has been shown to have disruptive effects in nematodes *in vitro*. In *Litomosoides carinii* and *A. suum* it increases glycogen synthase activity and, at the same time, inhibits glycogen phosphorylase in both species (Komuniecki and Saz 1982; Nelson and Saz 1982; Donahue, Masaracchia and Harris 1983). In this way, the mobilisation of glycogen is disrupted at a time when ATP levels are diminished as a result of the initial muscular paralysis induced by levamisole. The precise molecular site of action of levamisole on glycogen mobilisation has not been identified; it is thought not to act directly on these enzymes, and in *A. suum* it does not appear to affect Ca^{2+}-induced glycogenolysis. The evidence suggests that it is the cyclic AMP-controlled events that are affected, with levamisole having a postulated direct effect on adenylate cyclase.

In the glycolytic pathway in mammals the rate-limiting enzymes have been found to be the kinases: hexokinase, phosphofructokinase and pyruvate kinase. They have low activity in the tissues and they control reactions that are not readily reversible. Inhibitors of these enzymes, provided they are sufficiently selective, are effective antiparasitic agents. Because of the variety of terminal pathways present in parasites beyond the level of phosphoenolpyruvate, it might be expected that agents acting on enzyme steps *before* phosphoenolpyruvate would have a broader spectrum of activity than those acting after phosphoenolpyruvate. This is qualified, of course, by the level of selectivity or target specificity of the agent in question.

The fasciolicide clorsulon (MK-401), which structurally resembles 1,3-diphosphoglycerate, inhibits two enzymes of glycolysis in *F. hepatica*, phosphoglycerate kinase and phosphoglycerate mutase, causing inhibition of glucose utilisation and of end product formation (Schulman, Ostlind and Valentino 1982). This compound is effective *in vivo* for pharmacodynamic reasons. In the host, it binds to carbonic anhydrase in the erythrocytes. The blood-feeding parasites ingest the erythrocytes containing the drug and presumably liberate the active compound during digestion of the cells.

Organic antimony compounds inhibit enzymes by reacting with their sulfhydryl groups, many of which are sited on the molecule in locations essential for the enzyme's activity (e.g. the catalytic site); the glycolytic kinases are particularly sensitive to attack. 6-Phosphofructokinase from *S. mansoni* is considerably more sensitive to inhibition by stibophen or antimony potassium tartrate than the corresponding host enzyme, which is the basis for the effectiveness of these compounds against schistosomiasis (Bueding and Fisher 1966). The enzyme from adult filariae is also especially sensitive to stibophen (Saz and Dunbar 1975).

Organic arsenic compounds, which also react with sulfhydryl groups on enzymes, are effective against some protozoan groups. Melarsen and melarsen oxide inhibit pyruvate kinase in African trypanosomes *in vivo* and *in vitro* (see Wang 1984); other enzymes also inhibited by these compounds include the glycerol 3-phosphate oxidase complex, and pyruvate and 2-oxoglutarate decarboxylases. The vulnerability of pyruvate kinase in particular in African trypanosomes is probably due to its cytosolic location; the other susceptible enzymes are located within membrane-bound organelles (the glycosomes). In *L. m. mexicana*, hexokinase, pyruvate kinase, phosphofructokinase and malate dehydrogenase are all inhibited by melarsen oxide. Hexokinase in trypanosomatids is particulate and linked to the glycosomal membrane; in *T. cruzi* and *Leishmania* species its properties resemble those of yeast or *Neurospora* rather than the mammalian enzyme (Racagni and Machedo de Domenech 1983). These differences from the host may be worthy of further study if it can be shown that inhibition of glucose phosphorylation reduces the *in vivo* growth rate of these parasites.

Suramin inhibits a number of glycolytic enzymes in bloodstream African trypanosomes, including 6-phosphofructokinase and aldolase (Fairlamb, pers. comm. quoted by Meshnick 1984a). It also inhibits lactate dehydrogenase, malate dehydrogenase and malic enzyme from filarial worms, but suramin does not accumulate to concentrations within intact worms sufficient to inhibit glycolysis (Howells, Mendis and Bray 1983).

Some glycolytic enzymes in Protozoa have been shown to have different catalytic or structural properties from their mammalian counterparts, but specific inhibitors have not yet been found for clinical use. For example, phosphoenolpyruvate carboxykinase in many Protozoa resembles the enzyme from other groups of microorganisms in that it requires ATP as a substrate, compared with GTP in Metazoa (e.g. Broman, Knupfer, Ropars and Deshusses 1983). Lactate dehydrogenase in *P. falciparum* has different properties from the human enzymes (Van der Jagt, Hunsaker and Heidrich 1981). *E. histolytica* possesses a large number of glycolytic enzymes that differ quite significantly from those of their host, including phosphofructokinase that utilises pyrophosphate instead of ATP as substrate, and phosphoglycerate kinase that utilises guanine rather than adenine nucleotides. Two enzymes in *E. histolytica* have no counterpart in the host: pyruvate phosphate dikinase and PEP carboxyphosphotransferase both of which utilise pyrophosphate (see Reeves 1984a,b). These enzymes are 'unique receptors' and are potentially exploitable.

5.3.3 Unique subcellular organelles

Two groups of parasitic Protozoa possess unique types of subcellular organelle associated with energy metabolism. These are:

Glycosomes in the Kinetoplastids. These organelles have a single membrane and contain most of the glycolytic enzymes, plus enzymes of some other pathways (see Opperdoes, Misset and Hart 1984). The concentrations and close association of most of the glycolytic enzymes in these organelles in kinetoplastids is thought to facilitate the extremely high carbon flux through glycolysis; clearly an inhibitor or 'suicide' substrate entering such a concentrated system would have a rapid effect on glycolysis. In addition, the membrane surrounding the organelle might have interesting properties or transport systems that could be exploited. There are no examples at present of compounds acting specifically against glycosomes.

Hydrogenosomes in Trichomonads. These organelles appear to have a double limiting membrane, they contain large stores of glycogen and accommodate the pyruvate synthase complex and associated enzymes of the acetate-producing pathway. The reactions occurring in hydrogenosomes are important in the metabolic economy of the parasites, because resistant strains of *T. foetus*, grown in the presence of metronidazole, have been shown to grow considerably more slowly than normal organisms. This was attributed to loss of the oxidative catabolism of pyruvate and associated ATP synthesis in these strains (Cerkasovová, Cerkasov and Kulda 1984). Hydrogenosomes do not have cardiolipin in their membranes, nor do they contain DNA, which indicates that they are not homologous to mitochondria (Paltauf and Meingassner 1982; Wang and Wang 1985). Therefore, they may be susceptible to attack directed specifically at their membranes, for example.

5.3.4 Unique terminal pathways in energy metabolism

A number of unusual terminal pathways, not present in the hosts, have been identified in carbohydrate energy metabolism in parasites. They are therefore good targets for chemotherapy. These pathways diverge from glycolysis beyond the level of phosphoenolpyruvate. Agents directed at these terminal pathways are likely to be quite group- or even species-specific. Such pathways include:

Glycerol-3-phosphate oxidase complex in African trypanosomes
The bloodstream forms of African trypanosomes possess a number of unusual features associated with their unique pathways of aerobic glycolysis that are potentially exploitable by chemotherapy. Their terminal oxidase, ubiquinol:oxygen oxidoreductase, which is insensitive to cyanide, is unique and not found in mammals (Tielens and Hill 1985). It forms part of a mitochondrial enzyme complex that transfers electrons from glycerol-3-phosphate to oxygen.

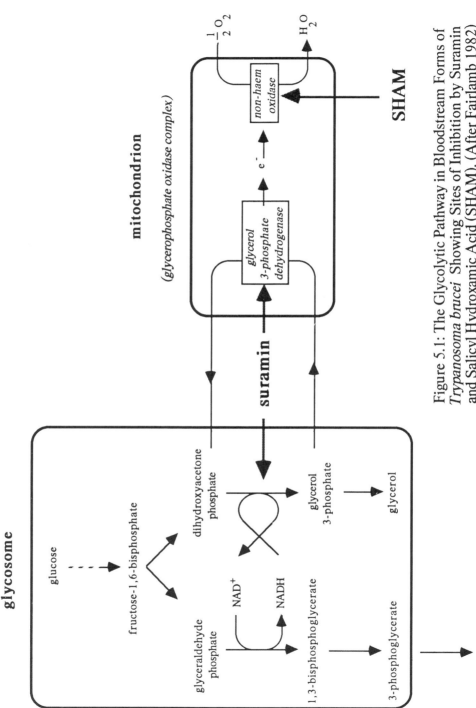

Figure 5.1: The Glycolytic Pathway in Bloodstream Forms of *Trypanosoma brucei* Showing Sites of Inhibition by Suramin and Salicyl Hydroxamic Acid (SHAM). (After Fairlamb 1982)

This complex also contains flavin-linked glycerol-3-phosphate dehydrogenase and ubiquinone. Inhibitors acting on any part of this system would affect aerobic glycolysis. Specific inhibitors for components of this pathway have been found (Figure 5.1). They include:

(i) Salicylhydroxamic acid (SHAM) and some related compounds that inhibit oxygen uptake non-competitively and appear to block electron transport between ubiquinone and the oxidase (see Meshnick 1984a, for review). SHAM does not kill trypanosomes unless the anaerobic glycerol-producing pathway is also inhibited.

(ii) Metal chelators, such as 8-hydroxyquinoline, benzhydroxamic acid derivatives and others inhibit oxygen uptake, though their precise site of action has not been identified (see Meshnick 1984a). Some chelators kill the trypanosomes *in vitro* without requiring simultaneous inhibition of the glycerol pathway.

(iii) Suramin inhibits the mitochondrial glycerol-3-phosphate dehydrogenase from *T. brucei* competitively with respect to glycerol-3-phosphate (Fairlamb 1981). Suramin also inhibits the glycosomal NAD^+-linked glycerol-3-phosphate dehydrogenase; this enzyme is *in vitro* more sensitive to inhibition than the mitochondrial one. The host enzyme is also less sensitive to inhibition than the glycosomal one. Parasites isolated from suramin-treated animals show reduced glycolytic flux and oxygen consumption (Fairlamb and Bowman 1980). An advantage of a compound such as this is that it inhibits both the aerobic and anaerobic pathways simultaneously.

It is clear that, although the glycerol-3-phosphate oxidase complex is a unique target in bloodstream trypanosomes, its inhibition is not sufficient to kill the organisms because the glycerol pathway provides an alternative means of reoxidising glycolytic NADH. Thus, an effective antitrypanosomal agent directed at this segment of glycolysis must inhibit both of the NAD^+-regenerating pathways.

Unusual pathways of pyruvate metabolism in some Protozoa
In *E. histolytica*, *G. lamblia* and the trichomonads, none of which possess mitochondria and all of which tolerate only low concentrations of oxygen, pyruvate is catabolised by unique pathways, as discussed previously (Chapter 2), that generate energy for the parasites and appear to be essential for their rapid growth.

Central to the function of these pathways is the enzyme pyruvate synthase (pyruvate:ferredoxin oxidoreductase) which oxidises pyruvate, transferring electrons to ferredoxin. Anaerobically, in *E. histolytica*, electrons are then transferred by undescribed carriers to NAD^+; aerobically, they are transferred by a flavin to oxygen to produce water (see Reeves 1984a). In *T. vaginalis*, the system also contains a hydrogenase, reducing protons to hydrogen anaerobically; aerobically, oxygen is reduced to hydrogen peroxide (see Yarlett, Gorrell, Marczak and Müller 1985). These systems are not found in the host and are therefore prime candidates for attack. They have also proved useful as parasite-specific reducing systems for activating certain classes of compounds that have suitable redox potentials to participate in the reactions. These unique pathways can therefore be exploited in two ways.

A known inhibitor of electron transport in this system is quinacrine, which is thought to bind to the flavin component in *E. histolytica* and *G. lamblia*, in which it suppresses respiration (see Weinbach 1981).

The reducing power of this system is essential for the activity of the nitroimidazoles, of which metronidazole is a commonly used example from clinical practice. These compounds have no antiparasitic activity until the nitro group is reduced by the parasites. The anion radicals that are synthesised intracellularly by the parasites have no effect on the host. In *T. vaginalis*, reduction of metronidazole is rapid anaerobically, much slower aerobically (Yarlett *et al.* 1985). The cytotoxic metabolites probably have a variety of targets, one of which is the parasite's nucleic acids (Moreno, Mason and Docampo 1984).

Fumarate reductase-linked terminal pathways in helminths
The majority of parasitic helminths, apart from the obligate aerobes or aerobic life cycle stages, have an active pathway branching from the glycolytic sequence at phosphoenolpyruvate and producing malate which is subsequently metabolised in the mitochondria. Malate may be metabolised in several ways. One way, from oxaloacetate to succinate, is a reversal of the TCA cycle reactions and is catalysed by the same enzymes with slightly different kinetic and regulatory properties from those of the host or of helminths with active TCA cycles. The second sequence leads via fumarate, and the enzyme complex called fumarate reductase (containing succinate dehydrogenase and electron transport components – see Chapter 2), to succinate, which may be excreted as an end product, or further metabolised to propionate or branched-chain volatile fatty acids, depending on the parasite species.

The first enzyme in the sequence generating malate is phosphoenolpyruvate carboxykinase. In helminths this enzyme predominantly operates in the CO_2-fixing direction, in contrast to the situation in vertebrates, where its major function is decarboxylation during gluconeogenesis. This enzyme would appear to be a strategic target for chemotherapy in helminths but to date no selective agent has been identified; the essential similarity of the parasite and host enzymes may preclude this.

These pathways, and especially the fumarate reductase reaction and its associated electron transport components, have for many years been considered to be prime targets for chemotherapy because they appeared to represent good examples of 'unique' pathways containing enzymes essential to the metabolic economy of the parasites. The results of work on these pathways have been disappointing from the chemotherapeutic point of view, for several reasons. First, the 'uniqueness' is questionable, because many of the same enzymes are found in the host, operating in the reverse direction. Second, the contributions of these pathways to the metabolic economy of some species of helminth have proved difficult to assess, partly because of intraspecific strain variation (see Chapter 6) and partly because it has become clear with some species that the TCA cycle plays a larger role in ATP synthesis than was originally supposed (see Chapter 2, Section 2.3.1). Third, good selective inhibitors of the fumarate reductase reaction or other enzymes of these pathways have not yet been found. Nonetheless, there are potential targets in these pathways, especially in the electron transport chain. Many helminths possess rhodoquinone instead of ubiquinone in a central position linking NADH dehydrogenase to succinate dehydrogenase or the terminal cytochromes (see Figure 2.23); they also have unusual types of cytochrome b and may also possess an alternative terminal oxidase (see Chapter 2; Köhler 1985; Barrett 1981). Complete inhibition of electron transport in most helminths would require either a specific inhibitor of the central rhodoquinone/cytochrome b complex, or several selective compounds each acting on a different branch of the chain.

Some of the anthelmintic benzimidazole compounds have been shown to inhibit the fumarate reductase complex. For example, thiabendazole inhibits fumarate reductase activity in isolated mitochondria of *A. suum* adults, acting at two sites in the electron transport chain – one between NADH dehydrogenase and rhodoquinone and the other between succinate dehydrogenase and the terminal oxidase (Köhler and Bachmann 1978). Three other compounds (levamisole, praziquantel and chloroquine) also inhibit electron transport at sites between NADH dehydrogenase and rhodoquinone. In all cases except chloroquine (which is not an effective anthelmintic) the drug concentrations required to exert significant inhibition were high and it is questionable whether this inhibition is responsible *in vivo* for the anthelmintic effects. For each of the above-mentioned anthelmintics, where they have *in vivo* activity, other essential worm processes are affected at much lower drug concentrations. These effects are therefore more likely to be primarily responsible for the worms' demise.

Several other enzymes of these terminal pathways in helminths are either unique or have different properties that are worth noting here, though there are no examples to date of compounds selectively attacking these targets. *A. suum* and *Toxocara canis* possess unique mitochondrial NAD-specific malic enzymes which have been found in no other group. In other helminths with NADP-specific malic enzymes mitochondrial NADPH-NAD transhydrogenases are present to supply NADH to drive the fumarate reductase reaction. The properties of the NAD-dependent enzyme may be worth investigating, since inhibition of this enzyme would inhibit the terminal anaerobic pathways utilising malate.

5.3.5 *Enzymes of oxidative metabolism*

Amongst helminth parasites, fully oxidative metabolism, with TCA cycle activity and 'classical' electron transport, is found in many eggs and larval stages of the life cycles, but is less common in adult stages, especially in large worms (see Köhler 1985 and Chapter 2). Many adult helminths possess some proportion of fully oxidative metabolism, contributing to ATP synthesis in parallel with the anaerobic pathways. There is some evidence that the aerobic pathways are restricted to certain locations around the periphery of larger helminths, or surrounding the gut in blood-feeding worms. But we do not know precisely how aerobic or anaerobic metabolism is regulated when (if) it occurs in the same tissue, and the apparent presence of branched electron transport chains complicates any proposed regulation schemes. Therefore, if aerobic energy metabolism is to be considered as a useful target for chemotherapy, we need to understand more about its contribution to the metabolic economy of each parasite species. Many inhibitors of aerobic energy metabolism are known from work on mammalian mitochondria. These include the uncouplers of oxidative phosphorylation, which increase mitochondrial permeability and abolish the transmembrane proton gradient, specific inhibitors of the electron transport system, and certain compounds that inhibit TCA cycle enzymes. To date there are no clinically useful inhibitors of aerobic metabolism that are specifically selective for helminth enzymes; those that are known are effective because they are distributed differently between host and parasite.

Substituted phenols and salicylanilides, such as nitroxynil, oxyclozanide, rafoxanide and closantel, all effective fasciolicides, are uncouplers of oxidative phosphorylation. These and similar compounds have the same effect on mammalian mitochondria *in vitro* (see Campbell and Montague 1981; Van den Bossche 1985) and, for most compounds, there is little significant difference in sensitivity between parasite and host mitochondria. When tested *in vivo*, however, the parasites succumb at lower doses than are required to affect the host. This may be due to the flukes' location in the liver, where the compounds may concentrate, and also, in the case of rafoxanide and closantel, to the fact that the drugs bind to serum albumin and do not enter the host's cells. Parasites ingesting blood, however, liberate the drugs from the albumin during digestion and are poisoned. This explains why both these compounds also kill *H. contortus*, a blood-feeding nematode in the abomasum of sheep. Parasites that rely on mitochondrial ATP generation, either anaerobic or aerobic, would be susceptible to uncouplers, provided that appropriate delivery systems that minimise toxicity to the host are found.

In parasitic Protozoa, fully oxidative metabolism is found – in the stages present in the mammalian host – predominantly in intracellular parasites, as discussed in Chapter 2. We know very little about the properties of enzymes of these pathways or the mechanisms controlling their activity, partly because of the technical difficulties associated with working on intracellular parasites. Therefore, few suitable specific targets have to date been satisfactorily demonstrated for these organisms.

Electron transport in parasitic Protozoa has, however, received some attention. It has been shown for *Eimeria* and for those trypanosomatids with complete oxidative metabolism that an important terminal oxidase is the cyanide-insensitive cytochrome o, which is not implicated in mammalian metabolism and is therefore a suitable chemotherapeutic target (see Fairlamb 1982). No specific inhibitors of this cytochrome have yet been reported. Electron transport (before the level of cytochrome o) in *Eimeria* is inhibited specifically by the 4-hydroxyquinolines, and also by metichlorpindol; these two groups of compounds act synergistically (see Wang 1984).

The electron transport systems of many helminth parasites contain rhodoquinone instead of ubiquinone (see Köhler 1985). Rhodoquinone, which has a more negative redox potential than ubiquinone, more readily donates electrons for the reduction of fumarate. Compounds that interact with rhodoquinone but not ubiquinone could be potential anthelmintics. Ubiquinone, too, has been identified as a useful target in some protozoan groups. In malarial parasites ubiquinone-8 is part of the respiratory chain, whereas its counterpart in mammals is ubiquinone-10. This may explain the activity of the naphthoquinones and quinolones against these parasites. Oxygen uptake and electron transport is disrupted by the hydroxynaphthoquinones in *Plasmodium* and also in *Eimeria* and *Theileria* (Hudson, Randall, Fry, Ginger, Hill, Latter, McHardy and Williams 1985). It should be noted that energy metabolism is not the only type of metabolism potentially disrupted in this case: other respiratory-chain linked pathways (e.g. *de novo* pyrimidine synthesis) may also be affected. Therefore, organisms that appear to synthesise ATP predominantly via glycolysis, such as many plasmodia, are still affected by electron transport inhibitors.

Oxidative phosphorylation is another potential target. We are unaware of any demonstrated specific targets in parasitic Protozoa, however, and little is known of the process in these organisms. In *Crithidia fasciculata*, oxidative phosphorylation by mitochondrial particles *in vitro* is inhibited by suramin. Therefore it is possible that some Protozoa with fully oxidative metabolism have a specific binding site for this compound (Roveri, de Cazzulo and Cazzulo 1982).

5.3.6 *Nucleic acid metabolism*

Nucleic acid synthesis is essential for rapidly growing cells and requires (in addition to appropriate energy supplies) the raw materials for synthesis, the nucleosides and nucleotides. Unlike energy metabolism, which must function continuously in active parasites, nucleic acid synthesis (especially DNA synthesis) may be cyclic or stage-specific. In Protozoa, which undergo rapid growth, DNA synthesis may be limited to specific phases of the life/cell cycle. Compounds affecting nucleic acid synthesis in Protozoa, therefore, may only be effective if presented to the parasites at certain specific stages of their life/cell cycles. They would not kill the parasites directly, but would interrupt or slow their rate of growth and multiplication. In helminths, however, DNA and RNA synthesis in adult worms would be largely restricted to the reproductive tissues, except in the case of continuously-growing cestodes. Compounds affecting nucleic acid synthesis would therefore inhibit egg production in all helminth groups and also affect growth of the strobila in cestodes.

As discussed in Chapter 3, in parasitic Protozoa and some parasitic helminths, purine and, for some groups, pyrimidine bases or their respective nucleosides are not synthesised *de novo* but are acquired from the host. They are then converted and interconverted to form the required nucleosides or nucleotides. Some of the pathways utilised, and the enzymes catalysing them, are different in the parasites from those in the host. In many instances, the host may not require or utilise salvage pathways at all and may supply its requirements by *de novo* synthesis. Therefore, these pathways in the parasite are a good source of useful targets for chemotherapy.

Suitable targets in purine or pyrimidine metabolism would include such processes as, for example, the mechanisms of uptake of the precursors from the host, or the specific enzymes of interconversion. Purine and pyrimidine uptake mechanisms in some helminth groups are different from those in mammalian cells and may be potentially exploitable. The latter could either be specifically inhibited or given 'suicide' substrates that are metabolised to compounds that derange subsequent reactions.

Purine metabolism

All groups of parasitic Protozoa and the majority of helminths appear to be unable to synthesise purines *de novo* and rely on host supplies of preformed bases or nucleosides. A network of reactions converts these into the nucleotides required for DNA or RNA synthesis and other processes (see Figure 3.6). These networks in parasites appear to be less complex than those present in mammalian cells and may therefore be more readily inhibited. It is not really possible to make general statements about purine salvage in the different groups of parasites, partly because little is known of the pathways in most helminths, and partly because the enzymes and bases required by different parasites, and available from different hosts, differ substantially. Even closely related parasites within similar taxonomic groups differ from each other; for example *T. vaginalis* and *T. foetus* salvage different bases and nucleosides and therefore have different enzymes for the conversion reactions (see Wang, Verham, Cheng, Rice and Wang 1984). Consequently, compounds directed at targets in this area of metabolism may be very specific for particular groups, or even species, of parasites.

Advantage can be taken of the fact that many purine salvage enzymes in parasites have slightly different substrate specificities from those of the host. This means that it should be possible to prepare stuctural analogues of the substrate that react with the parasite enzyme but not at all (or very slowly) with the host enzyme. Unfortunately, many of the analogues tested to date that have good antiparasitic activity are also quite toxic to the host (see Meshnick 1984b; El Kouni, Diop and Cha 1983). Unless protection methods can be devised for the host they cannot be used clinically.

Probably the most-studied enzyme for which analogues have been prepared as substrates is hypoxanthine-guanine phosphoribosyltransferase (HG-PRT). This, and related enzymes in Protozoa and (probably) helminths, metabolises a range of analogues not utilised by the host enzyme. The properties of HG-PRT also vary between different parasite groups (see Wang 1984). Allopurinol, a structural analogue of hypoxanthine, is effective against some kinetoplastids *in vivo*; it is a substrate for HG-PRT in these organisms, but not in the mammalian host. It is relatively non-toxic to the host and, indeed, has been used clinically for some time in the treatment of gout, because it inhibits xanthine oxidase. In *T. cruzi* and *Leishmania* species allopurinol is metabolised to adenosine nucleotide analogues by a series of reactions, and is incorporated into RNA. It has been suggested (Marr 1984) that the ribonucleotide monophosphates of pyrazolopyrimidines (such as allopurinol, thiopurinol) act in Protozoa, not by affecting the function of RNA, but by inhibiting the enzyme adenylosuccinate synthetase. In any case, the ultimate effect of these compounds is to reduce the growth rate of the sensitive parasites.

Many other purine analogues are in use experimentally against parasitic Protozoa (for review see, for example, Howells 1985). These include formycin B, an analogue of adenosine. Other purine-metabolising enzymes in parasitic Protozoa that may have different properties from those of the host include adenosine deaminase in plasmodia (Daddona, Wiesmann, Lambros, Kelley and Webster 1984), guanine deaminase in kinetoplastids (Kidder and Nolan 1981), and adenylsuccinate lyase in kinetoplastids (see Meshnick 1984b).

In helminths, purine salvage metabolism has been little studied, except in schistosomes and some filariae, where some important differences between host and parasite have been noted (see Senft and Crabtree 1983; Jaffe 1981; Jaffe and Chrin 1980). Schistosomes obtain considerable quantities of adenine nucleotides from their diet of red blood cells; the salvage pathways for these are very important in the worms' metabolic economy. The adenosine analogue tubercidin inhibits egg production and kills *S. mansoni in vitro* and *in vivo*. It interferes with adenosine metabolism and is metabolised via adenosine kinase to nucleotides which interfere with subsequent adenine nucleotide metabolism. Tubercidin is quite toxic to the host, but it has been used with success against *S. mansoni in vivo* in mice, when administered in combination with nitrobenzylthioinosine 5'-monophosphate. The latter compound inhibits nucleoside (including tubercidin) transport into the hosts' cells, but not those of the parasite, and thus reduces the toxic effect on the host (El Kouni *et al.* 1983). The combination is also effective against *S. japonicum* in mice (El Kouni, Knopf and Cha 1985). Thus, this combined treatment takes advantage of differences between host and parasite in both the substrate specificity of adenosine kinase and the mechanism of uptake of nucleosides into their cells.

In *D. immitis* and *O. volvulus* a particulate 5'-nucleotidase is required for conversion of 5'-mononucleotides to their corresponding nucleosides before uptake from the surrounding medium. The combination of this enzyme, acting on AMP to produce adenosine, and adenosine deaminase is thought to control the intracellular concentration of adenosine. Adenosine may have a regulatory role in parasites, as in mammals, by interaction with adenylate cyclase. A dithiocarbamate derivative of amoscanate, CGP 8065, inhibits this enzyme at low concentrations, while having no effect on the corresponding rat liver enzyme (Walter and Albiez 1985). Thus, this compound may affect not only intracellular adenosine as a purine nucleotide precursor, but also as a regulatory metabolite affecting, among other possible processes, adenylate cyclase-mediated events.

The effectiveness of substrate analogues such as allopurinol or tubercidin is determined not only by the potential of the parasite enzymes to interact with or metabolise them, but also initially by the ability of the parasite to take them up in sufficient concentration from its environment. For example, different strains of blood trypomastigotes of *T. cruzi* vary in their sensitivity to allopurinol. Although their drug-metabolising enzymes have similar activities, it has been shown that different strains remove the compound from the medium at different rates (Avila, Avila and Monzon 1984). Clearly, there is potential here for the development of resistant strains.

The presence of bypass enzymes in purine interconversion in some species creates problems for the use of analogues for chemotherapy. For example, *L. donovani* promastigotes grown under *in vitro* conditions in the presence of 4-aminopyrazolopyrimidine produced mutant strains whose growth was not inhibited by this compound (Iovannisci, Goebel, Allen, Kaur and Ullman 1984). It was shown that these strains were deficient (by genetic deletion) in adenine PRT for which 4-aminopyrazolopyrimidine is a substrate, yet their growth rate and rate of incorporation of adenine or hypoxanthine were similar to that of the wild type. In the mutant strains the enzymes adenine deaminase (for which 4-amino-pyrazolopyrimidine is *not* a substrate) and HG-PRT had taken over the role of adenine salvage. Whether this could become an effective mechanism for development of resistant strains *in vivo* is questionable because amastigotes of *L. donovani* do not have adenine deaminase activity and therefore depend on adenine PRT for adenine salvage. Nonetheless, the presence of potential alternative pathways could lead to the rapid development of resistant strains during clinical use. For this type of treatment to be effective in the long term, an additional inhibitor for adenine deaminase (in the example quoted above) would be required. Consequently, in determining a chemotherapeutic strategy, it is advantageous either to inhibit *all* the enzymes acting on a particular salvage substrate, or to inhibit a subsequent enzyme in the alternative pathway as well, probably by administering several compounds simultaneously. This type of approach, if it is to be rational, obviously requires a detailed understanding of the relevant pathways in the target species (including variation between strains), plus knowledge of appropriate uptake mechanisms for potential compounds, and the effects of the compounds on the host.

Some 'unique' enzymes of purine salvage have been found in parasitic Protozoa. For example, *G. lamblia* has a guanine PRT that does not recognise hypoxanthine, xanthine or adenine as substrates in contrast to the analogous enzyme in other organisms (see Wang 1984). *L. donovani* promastigotes have a separate xanthine PRT which is not present in the mammalian host (Tuttle and Krenitsky 1980). This parasite also possesses a 3′-nucleotidase on its external membrane which has a role in the uptake of 3′-nucleotides (the products of RNAse digestion). This enzyme has not been found in mammals. *Eimeria tenella* possesses a hypoxanthine, guanine, xanthine-PRT as a single enzyme, which is not known in any other organism (Wang and Simashkevich 1981).

Pyrimidine metabolism

Most groups of parasitic Protozoa are capable of pyrimidine biosynthesis *de novo*, except the trichomonads and *G. lamblia* (for review see Hammond and Gutteridge 1984). *T. gondii* and the kinetoplastids utilise *de novo* synthesis and also some salvage pathways. Evidence is accumulating that many helminths, too, are capable of *de novo* synthesis, though they have active salvage pathways as well. The *de novo* pathways in helminths may not be very active when salvageable pyrimidine nucleosides or nucleotides are present in their environments (see Barrett 1981). Mammalian hosts have active *de novo* and salvage metabolism. Because of the similarity in the *de novo* pathway between parasites and mammals, toxic pyrimidine analogues have not proved very selective in experimental chemotherapy. Nonetheless, the pyrimidine biosynthetic pathways remain potential targets, because it appears that humans can survive without *de novo* pathways, by utilising their salvage abilities, provided uridine is supplied (Kelley 1983). Since some experimental antimalarial compounds, for example, are directed at *de novo* pyrimidine synthesis, this ability may become important in chemotherapy (Hammond, Burchell and Pudney 1985).

Pyrimidine biosynthesis de novo. The pyrimidine *de novo* biosynthetic pathway has previously been discussed in Chapter 3. In parasitic Protozoa and in *S. mansoni* the pathways are qualitatively similar to those of their hosts, with no apparently unique enzymes to serve as potential targets. Some of the parasite enzymes have different properties from the host enzymes, however, though these differences have yet to be successfully exploited in chemotherapy. The most striking example in Protozoa is dihydroorotate oxidase (orotate reductase) which in most organisms is mitochondrial, membrane-bound, and is linked to the respiratory chain via ubiquinone. In the kinetoplastids, however, it is soluble, contains flavin, and appears to donate electrons directly to oxygen, forming H_2O_2 (see Meshnick 1984b). Therefore it is a potential target in this group. The link with the respiratory chain via ubiquinone has been exploited experimentally in plasmodia, where structural analogues of ubiquinone inhibit pyrimidine synthesis *in vitro* by preventing electron transfer to ubiquinone (see Hammond *et al.* 1985).

In helminths, *de novo* pyrimidine biosynthesis has been examined in detail only in *S. mansoni*, where some potentially exploitable differences between host and parasite have been found (see Iltzsch, Niedzwicki, Senft, Cha and El Kouni 1984). Orotate metabolism *in vitro* differed between parasite and mouse liver, and it was shown that the parasite orotate/uracil PRT was considerably more susceptible to inhibition *in vitro* by 5-azaorotic acid than the mouse liver enzyme. This enzyme activity is important in uracil salvage metabolism also. Different kinetic properties were also noted for orotidine 5'-phosphate decarboxylase and orotidine 5'-phosphate phosphohydrolase.

The intracellular distribution in kinetoplastids of some other enzymes of pyrimidine *de novo* biosynthesis is different from that in other organisms. Orotate PRT is particulate (glycosomal), whereas it is soluble in other organisms (see Hammond and Gutteridge 1984). This enzyme is different in *P. falciparum* too. It is physically separate from orotidine 5'-phosphate decarboxylase. In mammals the two enzymes are integrated into a bifunctional protein which has slightly different inhibition sensitivities from the plasmodial enzyme (Rathod and Reyes 1983).

Pyrimidine salvage. Pyrimidine salvage pathways, like the purine pathways, are characterised by a network of possible interconversions linking the bases and nucleotides of uracil and cytosine, but there is an important difference that may affect the vulnerability of pyrimidine pathways to chemotherapy. The pathways of salvage and metabolism of thymine or thymidine are completely separate from those of the other bases or nucleosides, except for one important linking enzyme. Since thymidine nucleotides are essential for DNA synthesis, attention to this area of metabolism in organisms without *de novo* pathways is potentially profitable. The enzyme linking thymidine and uridine metabolism is thymidylate synthetase an important enzyme that has a key role also in folate metabolism (see below). Thymidylate synthetase in kinetoplastids and plasmodia is present in physical association with dihydrofolate reductase as a bifunctional enzyme complex. In mammals, however, these two enzymes are separate. Such differences may be exploitable using specific inhibitors or substrate analogues for the bifunctional parasite enzyme. *T. foetus, T. vaginalis* and also *G. lamblia* do not possess either of these enzymes; they are unique amongst parasitic Protozoa in requiring exogenous thymidine, a vulnerable point in their metabolism (Wang and Cheng 1984a). The vital enzyme responsible for supplying TMP in the trichomonads is a thymidine phosphotransferase, with a contribution also from thymidine kinase in the case of *T. vaginalis* (Heyworth, Gutteridge and Ginger 1984). The phosphotransferase is clearly a potential point of attack. It is inhibited by guanosine or 5-fluorodeoxyuridine, which also slows the *in vitro* growth of the parasites (Wang 1984). It should be possible to design other suitable compounds to act at this site.

The enzyme uracil PRT is present in many protozoan and some helminth parasites but not in mammals (Hammond and Gutteridge 1984; Iltzsch *et al*. 1984). The analogue 5-fluorouracil inhibits this enzyme from *T. foetus* and also inhibits *in vitro* growth; incorporation of uracil, uridine and cytidine into nucleotides is inhibited, and the analogue is also incorporated into RNA (Wang, Verham, Tzeng, Aldritt and Cheng 1983). Analogues of uridine have also shown promise against some protozoan groups. 5-Fluorouracil and 5-fluorodeoxyuridine are also effective against the filarial worms *B. pahangi* and *D. immitis*, where they disrupt oogenesis and embryogenesis and inhibit the development of infective larvae, but do not kill micro- or macrofilariae (Howells, Tinsley, Devaney and Smith 1981). These analogues are too toxic to the host to be suitable for chemotherapy, but there remains the potential to find more selective compounds.

T. vaginalis, in contrast to all other Protozoa tested, including *T. foetus*, is able to convert neither its purine nor its pyrimidine ribonucleotides to the deoxyribonucleotides required for DNA synthesis, because it does not have a ribonucleotide diphosphate reductase activity (Wang and Cheng 1984b). In this parasite, the deoxyribonucleotides are supplied by a deoxyribonucleotide phosphotransferase acting on salvaged deoxyribonucleosides. This unique situation should be exploitable.

5.3.7 *Nucleic acid synthesis*

The early stages of nucleic acid biosynthesis are susceptible to structural analogues of substrates (i.e. antimetabolites), for the enzymes that synthesise pyrimidine and purine nucleotides, as discussed above. Many compounds also directly affect the synthesis of DNA and RNA. These are used in antimicrobial and anticancer chemotherapy, as well as against parasites, but are limited by their toxicity to the host. With a single known exception (kinetoplast DNA), unique targets in DNA or RNA synthesis have not been identified in parasites and efficacy is due either to more rapid uptake of the compound by the parasite, to a more sensitive parasite target, or to inability of the parasite to restore normal activity after treatment.

DNA synthesis can be inhibited by compounds that bind directly to, or intercalate into, the DNA molecule, preventing the helix from unwinding and thus preventing transcription or replication, or both. The diamidines (e.g. pentamidine, diaminazine aceturate [Berenil]) kill African trypanosomes at low concentrations *in vitro*. They interact directly with the phosphate groups of the nucleotides and insert into the helix. Trypanosomes are particularly susceptible because they have an energy-dependent rapid uptake of these compounds (see Meshnick 1984a). Chloroquine, mepacrine and primaquine all intercalate with DNA in many organisms, but this may not necessarily be the mechanism of their antiplasmodial action (see Desjardins and Trenholme 1984). The antitrypanosomal compounds ethidium and quinapyramine inhibit DNA or RNA synthesis probably by intercalation and physical disruption of replication.

Kinetoplast DNA (kDNA) is the kinetoplastid equivalent of mitochondrial DNA and is characterised by its abundance and its structural organisation into minicircles and maxicircles. Maxicircles contain the equivalent of mitochondrial DNA; the function of minicircle DNA, which is heterogeneous and rapidly evolving, is not understood. Because the maxicircles and minicircles are present in a concatenated mass in the parasites, and because they must be precisely divided up into daughter kinetoplasts at cell division, it is likely that the processes of unravelling and replication are both different from those in other organisms, and also precisely controlled. Therefore, elucidation of this replication mechanism may reveal unique processes or enzymes that may be susceptible to attack. In addition, kDNA has no histone covering and is therefore more exposed for potential attack than nuclear DNA. It has been observed that the cationic antitrypanosomal drugs (e.g. Berenil, pentamidine) bind to a greater extent to kDNA, causing cross-linking, than to nuclear DNA (see Albert 1985; Fairlamb 1982), but it has not been demonstrated that this is their primary site of action. Functional kDNA is required by most kinetoplastids for synthesis of some mitochondrial proteins, though it has been reported that *T. evansi* and *T. equiperdum* do not have a complete kDNA network (Fairlamb 1982). Bloodstream forms of *T. brucei* do not require significant expression of kDNA until mitochondrial elaboration commences in the short stumpy forms and therefore may be less susceptible to inhibitors of kDNA expression than those kinetoplastids with fully functional mitochondria. All kinetoplastids, however, need to replicate their kDNA before cell division and would be susceptible to compounds inhibiting replication.

Direct destruction of DNA, as opposed to physical hindrance of replication, has been reported for some antiprotozoal compounds. Such drugs, of course, are dangerous to the host too and selectivity is difficult to achieve. Nifurtimox and benznidazole cause strand breakage in kDNA and nuclear DNA of *T. cruzi in vitro* (Goijman, Frasch and Stoppani 1985). In the case of nifurtimox, the damage is a result of redox cycling that generates the nitrofuran anion plus the superoxide radical. This parasite has inadequate defences against this radical (see below) and the net result is an inhibition of DNA, RNA and protein biosynthesis. The antitrypanosomal action of bleomycin may also be mediated by oxygen radical-induced cleavage of the nuclear DNA, causing strand breakage; *in vitro* its effect requires the presence of oxygen and Fe^{2+} (see Meshnick 1984a). Two significant factors contributing to the efficacy of bleomycin are that it physically intercalates into the DNA before causing strand cleavage, thus also hindering repair, and that the products of oxidative cleavage, which are released, are highly cytotoxic, especially to cancer cells (see Albert 1985). Another antiprotozoal compound directly damaging DNA via radical metabolites is metronidazole, which is discussed in more detail below. The basis for its selectivity is its specific metabolism by the susceptible parasites to toxic radical metabolites.

Agents damaging DNA or hindering its synthesis may also have a similar effect on RNA, thus also interrupting translation. Ethidium and bleomycin, however, affect only DNA. Quinapyramine, an intercalating agent, inhibits RNA synthesis but not DNA synthesis in *Crithidia oncopelti* and also causes aggregation of ribosomes. Similar ribosomal aggregation has also been observed in trypanosomes from quinapyramine-treated hosts (see Meshnick 1984a).

The antischistosomal compound hycanthone has been reported to inhibit RNA and, later, DNA synthesis in *S. mansoni*, where it interferes with uridine incorporation into RNA (Mattoccia, Lelli and Cioli 1981). Although it also inhibits RNA synthesis in mammalian cells, this effect is reversed when the drug is removed; in the parasites, on the other hand, the effect is irreversible. Parasite strains resistant to hycanthone also show a reversible effect. The precise molecular site of action for this compound has not yet been identified.

5.3.8 *Targets in growth and development*

Processes in growth and development that are potential targets for chemotherapy include parasite-specific growth requirements (nutrients) and their uptake mechanisms, protein synthesis and other anabolic pathways, pathways providing cofactors or substrates for anabolic metabolism, and mechanisms controlling development. We have much to learn about all these aspects of parasite metabolism, but some possible targets have been identified.

Specialised nutritional requirements
Since parasites are dependent on their hosts for all their requirements for growth and since they lack many pathways for synthesising or processing essential substrates, this area of their metabolism is well worth investigating. Considerable work is required on this aspect of host-parasite relations and it is not possible to generalise about groups or species of parasites. Therefore we will illustrate this section with selected examples only.

The majority of parasites of mammals require a source of carbohydrate to supply their energy metabolism; this is especially true for the larger helminths that have little aerobic metabolism and are unable to utilise lipids or amino acids. In general, the carbohydrate substrate for helminths and many Protozoa is glucose, though some have a well-developed capacity to utilise glycerol also. Mechanisms of uptake for these or similar vital substrates for energy metabolism are worthy of investigation as potential targets for chemotherapy. Glucose uptake in cestodes appears to be largely by active transport, while in other helminth groups it occurs by a variety of methods, depending on the taxonomic group and nature of the habitat (see Barrett 1981; Uglem, Lewis and Larson 1985). Although the transport mechanisms have been characterised in detail in several species, the results have been disappointing from the point of view of chemotherapy, because they show no fundamental differences from transport mechanisms employed by host cells. Therefore, uptake mechanisms for molecules participating in processes less universal than energy metabolism may be better potential targets for chemotherapy. Most of the available examples in this category are found amongst the protozoan parasites.

A number of antiparasitic compounds have been shown to affect the uptake of glucose by helminth parasites. These include some of the benzimidazole anthelmintics, praziquantel, closantel and diamfenetide (see Behm and Bryant 1985; Andrews and Thomas 1979; Van den Bossche and Verhoeven 1982; Edwards, Campbell, Sheers, Moore and Montague 1981). But, in each case, impairment of glucose transport was either not severe or likely to be a secondary effect resulting from either impairment of energy metabolism (in cases where glucose transport is an active process) or physical damage to the tegumental membrane.

Filarial parasites possess specific retinoid-binding proteins for the uptake and transport of vitamin A, which appears to have a role as an intermediate lipid carrier in glycoprotein synthesis (Sani and Comley 1985; Comley and Jaffe 1983). These proteins have different properties from the host's and may be useful chemotherapeutic targets if selective synthetic retinoids can be found that bind tightly and disrupt vitamin A transport. Alternatively, specific antibodies to these 'receptors', which may be located on the external cell surface, could also be effective.

Kinetoplastids are unable to synthesise haem because they lack the *cytosolic* enzymes of the porphyrin pathway, though *T. cruzi* possesses the *mitochondrial* enzymes of the pathway (Salzman, Stella, Wider de Xifra, Batlle, Docampo and Stoppani 1982) and the final enzyme in the pathway, mitochondrial ferrochelatase is inducible in *L. m. amazonensis* (Chang and Chang 1985). Haem or protoporphyrin IX is required by these parasites for the synthesis of cytochromes, catalase and peroxidase and must be obtained by uptake from the host. Since the host can synthesise haem *de novo*, inhibitors of haem or protoporphyrin IX uptake by the parasites may be useful for chemotherapy, though none is at present identified. It is of interest in this context that the intracellular stages of *L. m. amazonensis* do not utilise the haem synthesised by their host macrophages, but remove haem independently from the medium. They may therefore be susceptible to this type of attack, even though they are intracellular. Other intracellular species – many of which require cytochromes for their energy metabolism – as well as extracellular species may also be vulnerable.

Parasites also need non-haem iron for growth, since it is required for iron-dependent enzymes such as ribonucleoside diphosphate reductase some oxidases and hydroxylases. Iron chelators such as desferrioxamine have been shown to inhibit *in vitro* growth of *P. falciparum* and intracellular *T. cruzi* (Raventos-Suarez, Pollack and Hagel 1982; Loo and Lalonde 1984). *In vivo* it has been shown that depletion of the host's intracellular iron stores reduces the pathogenicity of *T. cruzi*, but that normally the host responds to intracellular depletion of iron by increasing its rate of transfer into these cells, thus permitting the parasites to grow more rapidly. If this host response could be prevented while at the same time depleting circulating iron, parasite growth would be severely inhibited. But such treatment would probably not be curative if used alone, because the host also requires non-haem iron and would be affected by long-term interference with the supply of iron.

Plasmodial parasites in the red blood cell consume haemoglobin, which supplies them with essential amino acids. When the host cell's haemoglobin is modified by agents (e.g. dibromoaspirin or *bis* (dibromosalicyl) diesters) that cause acetylation and cross-linking of the molecule, the plasmodial cathepsin B and proteases are unable to digest it. Modification of haemoglobin by these compounds, which are specific for haemoglobin, does not interfere with its oxygen-transporting properties, and so it is potentially useful *in vivo. In vitro* the presence of these haemoglobin modifying agents is toxic to *P. falciparum* (Geary, Delaney, Klotz and Jensen 1983).

Malarial parasites of mammals actively concentrate Ca^{2+} from the host cell. Inhibitors preventing Ca^{2+} uptake by the red blood cell cause the death of the parasite within, without disturbing the plasma Ca^{2+} concentration (McAlister and Mishra 1983). Such inhibitors include vanadate, procaine-HCl and the ionophore A23187. This and the previous example are interesting in that the potential chemotherapy takes advantage of a function essential to the parasite, but is directed at the *host* and thus only indirectly at the parasites.

Differences in amino acid transport have been noted between parasites and their hosts. For example, *Eimeria* species have been shown to possess a unique transporter for thiamine that is very sensitive to inhibition by the thiamine analogue amprolium, which is effective as an anticoccidial *in vivo* (James 1980). The host thiamine transporter is considerably less sensitive to this analogue. It has been shown that *Leishmania* promastigotes have transport mechanisms for neutral amino acids that are different from their hosts (Bonay and Cohen 1983). There is no doubt that other suitably exploitable examples will come to light as research in this field proceeds.

Protein synthesis

A number of stages in protein synthesis are susceptible to chemotherapy. The metabolism by parasites of substrate analogues of purines and pyrimidines to generate inhibitory nucleotides or faulty RNA or DNA, as discussed above, will potentially affect protein synthesis. This has been demonstrated, for example, for the purine analogue formycin B, which is metabolised by *L. mexicana in vitro* to nucleotides that inhibit RNA and protein synthesis within 30 minutes (Nolan, Berman and Giri 1984). Although messenger RNA was implicated in this case, derangement of transfer and ribosomal RNAs could also occur.

The first step in the process of protein synthesis is the binding of tRNAs by amino acids, catalysed by specific aminoacyl-tRNA synthetases. In higher eukaryotes, some of these enzymes are found in high molecular weight complexes specific for different amino acid substrates. Some of these enzymes are under regulatory control by phosphorylation-dephosphorylation. In preparations from *A. suum* leucyl- and isoleucyl-tRNA synthetases were inhibited by low concentrations of CGP 8065, the dithiocarbamate derivative of amoscanate (a broad-spectrum anthelmintic) (Walter and Ossikovski 1985). If this inhibition also occurs *in vivo*, it could explain some of the observed ultrastructural damage and the apparent long-term inhibition of macromolecular synthesis that occurs in helminths after treatment with amoscanate.

Direct inhibition of the translation process at the ribosomal level may be brought about by agents interacting with ribosomes. Bacterial ribosomes have different physical properties from those of eukaryotes, which make them susceptible to many common antibiotics (e.g. chloramphenicol, streptomycin). These compounds may also affect protein synthesis in the mitochondria of eukaryotes if they gain entry. *Unique* targets for direct inhibition have not been found in parasites, but certain inhibitors of protein synthesis appear to be effective because protein synthesis may be more crucial at certain phases of the life cycle, or because their metabolic machinery for recovery from inhibition is less efficient in parasites than in their hosts.

Emetine, which has been used for many years against *E. histolytica*, inhibits peptidyl-tRNA transfer within the ribosomes in eukaryotes. *E. histolytica* is more susceptible than its host because it recovers more slowly (see Albert 1985; Entner 1979). The growth of *P. falciparum* trophozoites *in vitro*, and their incorporation of isoleucine, is inhibited by chloramphenicol, erythromycin and tetracycline, all agents that inhibit mitochondrial protein synthesis (Blum, Yayon, Friedman and Ginsburg 1984). This inhibition only occurs at the trophozoite stage of the life cycle. It is interesting that these parasites require mitochondrial protein synthesis for growth even though they do not appear to have a fully oxidative mitochondrial energy metabolism. Other agents inhibiting protein synthesis in Protozoa include benznidazole, nifurtimox and SQ 18,506 (a nitrofuran) which all affect *T. cruzi in vitro*, though their molecular sites of action are unknown (Polak and Richle 1978; Goijman *et al.* 1985; Gugliotta, Tanowitz, Wittner and Soeiro 1980).

Anabolic metabolism
In parasites, the majority of anabolic pathways, their substrates, products, interactions and control mechanisms have received little attention from researchers to date. Consequently, few potential drug targets have been identified in parasite anabolic metabolism, though we know that there are likely to be differences between parasites and their hosts because of the parasites' generally limited biosynthetic abilities. This section is of necessity incomplete, but some useful areas of investigation can be discussed.

Lipid biosynthesis. Lipids, phospholipids and sterols are essential for growth and the maintenance of membranes in all organisms. The limiting or external membranes of parasites are especially important, since they must both protect the parasite from host attack and, at the same time, permit and perhaps control the uptake of nutrients across the membrane. The latter function is especially important for parasites without an internal gut. Therefore, interference with lipid uptake or biosynthesis could be a useful strategy in chemotherapy. The complex pathways of lipid biosynthesis in important parasites have received little systematic attention in the past, but are beginning to recieve some close scrutiny.

The majority of parasites appear to rely on host-supplied precursors for lipid synthesis or modification and therefore may have specific uptake mechanisms for these compounds. They may also rely on certain synthetic pathways that utilise these precursors to a greater extent than their hosts, and possibly employ different regulatory processes. Several unique pathways or enzymes have been identified in some parasitic Protozoa. For example, acetate units for lipid synthesis in bloodstream and culture *T. brucei* are preferentially derived from threonine by the action of threonine dehydrogenase and glycine acetyltransferase This pathway does not occur in the host. Disulfiram inhibits threonine dehydrogenase and kills trypanosomes *in vitro*, but is not active *in vivo*, possibly because the parasites do not absorb it under these conditions (see Meshnick 1984b). In *P. falciparum* and *P. knowlesi*, certain analogues of choline and ethanolamine inhibit *in vitro* parasite growth. 2-Aminobutanol was incorporated by *P. knowlesi* into an unnatural phospholipid which accumulated and presumably had no function (Vial, Thuet, Ancelin, Philippot and Chavis 1984). Two active enzymes have been identified in *P. knowlesi* that are not present in the host's red blood cells. They are phosphatidylserine decarboxylase and phosphatidylethanolamine methyltransferase (Vial, Thuet, Broussal and Philippot 1982).

Parasite membranes differ in composition from those of their hosts. For example, membranes of plasmodial parasites have higher concentrations of unesterified fatty acids, triacylglycerols, 1,2-diacylglycerols, diacylphosphatidylethanolamine and phosphatidylinositol and lower concentrations of cholesterol, phosphatidylserine and sphingomyelin than their host's membranes (Sherman 1983). This not only means that parasite membranes have different properties from those of their hosts, but also that parasites possess a different (quantitatively, at least) synthetic machinery for supplying membrane constituents, or specific uptake processes for obtaining precursors from the host. Specialised external membranes are present in some parasites. Adult *S. mansoni* possess a double outer membrane which develops a few days after penetration of the host. This membrane is very rich in lipids, it is constantly renewed, and it has a vital function as the interface between the parasite and the host (Vial, Torpier, Ancelin and Capron 1985).

Sterol biochemistry in *L. m. mexicana* is different from that of the host and resembles that of fungi (Goad, Holz and Beach 1985). The antimycotic agent, ketoconazole, inhibits growth of this parasite *in vitro* and causes the accumulation of cholesterol and other sterols. It is thought to act by inhibiting the cytochrome P-450-dependent 14α-demethylation of lanosterol. Ketoconazole is also effective *in vitro* against intracellular *T. cruzi* and *P. falciparum* (McCabe, Remington and Araujo 1984; Pfaller and Krogstad 1983). Since agents interfering with sterol biosynthesis have been very useful in treating fungi of medical importance, there is potential here for investigating similar compounds in the treatment of Protozoa.

Parasitic helminths (and their free-living relatives) are unable to synthesise steroids *de novo*, though some have been shown to synthesise isoprenoids (e.g. quinones, dolichols), and to modify externally supplied cholesterol (see Barrett 1981; Comley 1985). Some unique pathways have been identified. Helminths have the ability to synthesise and excrete ecdysteroids, juvenile hormone and farnesol, compounds that play essential roles as hormones in the development of insects. Do these compounds have similar important functions in helminths? If so, it may be possible to exploit them in chemotherapy using specific inhibitors of synthesis or 'antihormones' (see Mercer 1985; Waller and Lacey 1985). Helminths also appear to be unique (amongst those eukaryotes that have been examined) in that they can directly form rhodoquinone-9 from ubiquinone-9, a reaction that may lend itself to inhibition. Rhodoquinones are not found in vertebrates and their function in helminth electron transport may be sufficiently different from ubiquinone to be selectively inhibited. Dolichol kinase, a rate-limiting enzyme in the synthesis of dolichyl monophosphate (which participates in glycosyl transfer for glycoprotein synthesis), has been shown in *O. volvulus* and *A. suum* to have different regulatory properties from the mammalian enzyme, in that Ca^{2+}-calmodulin is not a regulator (Walter, Ossikovski and Albiez 1985). No doubt other interesting differences in synthetic pathways such as these will be demonstrated upon further investigation.

Folate biosynthesis and metabolism. Dihydrofolate is an essential requirement for growth and reproduction in all organisms because derivatives of its reduced form, tetrahydrofolate (THF), are essential cofactors for many methylation and other carbon transfer reactions. For example, 5,10-methylenetetrahydrofolate (5,10-MTHF) is a substrate for thymidylate synthetase in the conversion of dUMP to dTMP, which is required for DNA synthesis in parasites unable either to synthesise thymine or to salvage thymidine. If parasites employ *de novo* purine synthesis as may occur in the filariae (see Barrett 1983), they would require 5,10-MTHF and 10-formylTHF as substrates. Therefore, interruption of folate supply or metabolism will inhibit growth and reproduction in parasites.

Intracellular sporozoa and pathogenic bacteria synthesise dihydrofolate *de novo* from the precursors GTP, *p*-aminobenzoate and glutamate (see Figure 5.2), whereas their mammalian hosts do not have this pathway; they recover folate from their diet and reduce it to dihydrofolate directly. This substantial difference in the source of dihydrofolate has been useful in chemotherapy because analogues of *p*-aminobenzoate competitively inhibit the enzyme dihydropteroate synthase thus diminishing the rate of dihydrofolate synthesis.

Sulphonamides and sulphones (such as sulphathiazole, sulphaguanidine, sulphanilamide, sulphadoxine, dapsone) inhibit this enzyme in *Plasmodium* species, and also in other intracellular sporozoans, *Toxoplasma* and *Eimeria* (see Wang 1984). There is some evidence that the properties of dehydropteroate synthase in *Plasmodium* are different from those in pathogenic bacteria.

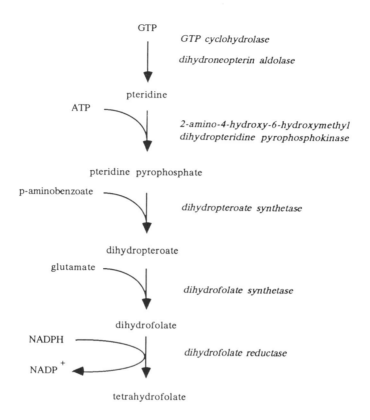

Figure 5.2: The *de novo* Synthesis of Tetrahydrofolate

Another important chemotherapeutic target in folic acid metabolism is the enzyme dihydrofolate reductase (DHFR) which converts dihydrofolate to tetrahydrofolate. A well-known inhibitor of this enzyme is the pteridine derivative methotrexate, a substrate analogue which is used in cancer chemotherapy. This drug is only moderately selective, which causes toxicity problems to the host, and it is not generally taken up by bacteria or most Protozoa, which do not have specific transport mechanisms (see McCormack 1981; Albert 1985). But it inhibits *in vitro* growth of *Leishmania* species (Scott, Coombs and Sanderson 1984) and is a potent inhibitor of DHFR from *P. berghei* (Pattanakitsakul and Ruenwongsa 1984). As mentioned earlier, this enzyme in plasmodia and kinetoplastids is different in that it forms a bifunctional complex with thymidylate synthase. Therefore it may be susceptible to more selective substrate analogues.

Pyrimethamine, a simplified derivative of the 2,4-diaminopyrimidinyl moiety of methotrexate, is a powerful antimalarial that is highly selective and widely used in prophylaxis. Plasmodial DHFR is a thousand times more sensitive to inhibition by this compound than the mammalian enzyme (see Albert 1985). Leishmanial DHFR is less sensitive (Scott *et al.* 1984). Other inhibitors of DHFR in use against plasmodia include trimethoprim, proguanil (which is paludrine, metabolised to cycloguanil) and clociguanil. Trimethoprim is also active *in vitro* against trypanosomes but does not work *in viva,* possibly because of the high concentrations of dihydrofolate found in trypanosomes (see Meshnick 1984b). Interestingly, suramin, an unrelated compound, also inhibits DHFR from *T. b. brucei.* The enzyme from the filarial worm *O. volvulus* appears to be unique amongst the metazoan examples tested, in that it is particularly sensitive to suramin (Jaffe 1972).

   Combinations of agents that attack both of these enzyme sites in folate metabolism act *synergistically* because greater pathway inhibition is achieved by inhibiting two sequential sites in a reaction series. It is important when designing such drug combinations that the components have similar half-lives *in vivo* in order to achieve maximum effect and to reduce the rate of development of resistance to the individual components. Two important current antimalarial treatments employing this principle are Maloprim, with pyrimethamine and dapsone, and Fansidar, with pyrimethamine and sulfadoxine. Toxoplasmosis has also been effectively treated with a combination of high concentrations of pyrimethamine plus sulphonamide; the host was protected from pyrimethamine inhibition of its DHFR by supplementation with leucovorin (reduced folate), which cannot be utilised by *Toxoplasma* (McCormack 1981).

   Folate metabolism in helminths has been investigated in detail only in the filariae (see Barrett 1983), where a number of possible chemotherapeutic targets have been identified. These worms do not appear to synthesise dihydrofolate *de novo,* but obtain 5-methyltetrahydrofolate (5-CH$_3$THF) as their major source of folate from the host. The pathways of folate metabolism identified in adult filariae are illustrated in Figure 5.3. In general, the folate-metabolising enzymes in adult filariae have similar properties to their mammalian or mosquito counterparts (Jaffe and Chrin 1980; Jaffe, Chrin and Smith 1980), with several important differences. One of these is that filariae are able to oxidise 5-CH$_3$THF to 5,10-MTHF because the responsible enzyme, 5,10-MTHF reductase, in filariae, is able to catalyse the oxidation reaction (reaction 4 in Figure 5.3), whereas the host enzyme catalyses only the reduction reaction. This enzyme, which is a flavoprotein, would therefore be a potential target and, indeed, has been shown to be sensitive to inhibition by menoctone (an analogue of ubiquinone) which presumably interrupts electron transport; *in vitro,* with 5-CH$_3$THF supplied in the medium, movement and production of microfilariae is inhibited by menoctone (Jaffe 1980). Another difference between host and parasite is that the enzyme formylTHF dehydrogenase has considerably higher activity in the parasites; it is sensitive *in vitro* to inhibition by the filaricides diethylcarbamazine and suramin (reaction 9). Diethylcarbamazine also inhibits glutamate formiminotransferase (reaction 11) and methylenetetrahydrofolate dehydrogenase *in vitro* (reaction 5), but it has not been shown whether these inhibitions contribute significantly to the anthelmintic activity of this compound.

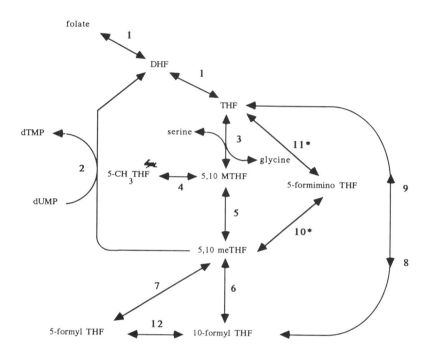

Enzymes	Abbreviations
1. dihydrofolate reductase 2. thymidylate synthetase 3. glycine hydroxymethyl transferase 4. 5,10-methylene tetrahydrofolate reductase 5. methylene tetrahydrofolate dehydrogenase 6. methylene tetrahydrofolate cyclohydrolase 7. 5-formyltetrahydrofolate cycloligase 8. formate-tetrahydrofolate ligase 9. formyltetrahydrofolate dehydrogenase 10. formiminotetrahydrofolate cyclodeaminase 11. glutamate formiminotransferase 12. proposed 5-formyl, 10-formyltetrahydrofolate mutase	DHF: 7,8-dihydrofolate THF: 5,6,7,8-tetrahydrofolate 5,10-MTHF: 5,10-methylene tetrahydrofolate 5-CH$_3$ THF: 5-methyl tetrahydrofolate 5,10 meTHF: 5,10-methenyltetrahydrofolate * bifunctional protein ⟿ derived from host

Figure 5.3: Pathways of Folate Metabolism in Adult Filariae

Special synthetic processes. Chitin, a polymer of N-acetylglucosamine with $\beta(1\to4)$ linkages, is found in many helminth eggs (Wharton 1983), in the protective sheath of microfilariae (Fuhrman and Piessens 1985) and in the cyst walls of *Entamoeba* spp. (Avron, Deutsch and Mirelman 1982). It is not found in mammals, and chitin synthesis therefore is a potential target for chemotherapy in parasites. Since a range of compounds affecting chitin synthesis is available from insect studies, it is logical to test their effects on parasites. Inhibitors such as polyoxin D and nikkomycin (both analogues of UDP-N-acetylglucosamine) prevent cyst formation *in vitro* by trophozoites of *E. invadens* (Avron *et al.* 1982). In *B. malayi*, diflubenzuron, a specific inhibitor of chitin synthesis, caused gravid females *in vitro* to produce microfilariae with deformed sheaths (Fuhrman and Piessens 1985). If compounds such as these are effective *in vivo*, they would affect transmission of the parasites. Thus, chitin synthesis inhibitors could be valuable additions to the chemotherapeutic arsenal.

The specialised processes by which parasites synthesise their egg shells or cyst walls may also prove to include reactions that do not occur in the host and inhibition of which would interrupt the parasite's life cycle. For example, many helminth eggs contain quinone-tanned proteins that are formed by the action of phenolases on phenols, giving quinones that are responsible for cross-linking proteins. In *S. mansoni*, inhibitors of phenol oxidases, including disulphiram, prevent egg production *in vivo* (Seed and Bennett 1980). With further work, other specialised processes in egg shell or cyst wall production in parasites may be identified that will be susceptible to chemotherapy.

5.3.9 *Mechanisms of control and regulation*

A number of regulatory components or processes have been identified that control or regulate metabolism, growth and development in parasites. These include cyclic AMP and the processes it affects, protein kinases, calmodulin and other factors controlling or controlled by Ca^{2+} levels, polyamine biosynthesis and function, and methylation and other reactions in pathways that are important for parasite growth. In this section we will discuss some examples where potentially exploitable differences between host and parasite have been identified.

Cyclic AMP and protein kinases
Cyclic AMP (cAMP) plays a central part in metabolic regulation. It causes the activation (by phosphorylation) of some regulatory enzymes, and also increases phosphorylation of some membrane proteins, which may be important in the regulation of neuromuscular activity. The intracellular concentration of cAMP is determined by the relative activities of adenylate cyclase and cAMP-phosphodiesterase. Inhibition of either of these enzymes would seriously disrupt the control of cAMP-dependent cellular processes.

The activity of adenylate cyclase in *F. hepatica* is unusually high and the enzyme is important in the regulation of carbohydrate metabolism and motility in trematodes (see Mansour 1984).   Adenylate cyclase in this and other helminth parasites is sensitive to activation by 5-hydroxytryptamine (5-HT), but not to epinephrine, which is an activator in mammalian tissues.   The 5-HT receptors associated with liver fluke adenylate cyclase have different properties (with respect to the binding of agonists and antagonists of 5-HT) from their mammalian counterparts.   In *S. mansoni* in mice the activity of 5-HT-activated adenylate cyclase increases after infection and during development and maturation, which suggests a requirement for increased cAMP synthesis during growth.   In *A. suum* muscle 5-HT also stimulates cAMP synthesis but does not increase motility in this parasite.   This enzyme in helminths is therefore a potential target for chemotherapy.   We have discussed above the possibility that levamisole, by inhibiting the adenylate cyclase of *A. suum* muscle, inhibits cAMP-mediated (but not Ca^{2+}-mediated) glycogenolysis.

cAMP-phosphodiesterase may also be an important site for chemotherapy, though this enzyme from parasites has received little attention from researchers. Amoscanate, a broad-spectrum anthelmintic, inhibits cAMP-phosphodiesterase in *S. mansoni* and *O. volvulus*, causing accumulation of cAMP which then activates the glycogenolytic cascade (causing depletion of glycogen) and probably other cAMP-dependent processes (Walter and Albiez 1984).

In parasitic Protozoa, too, cAMP appears to affect the regulation of development.   It inhibits cell division *in vitro* in trypanosomatids, and in *P. falciparum* it inhibits asexual multiplication and gametocyte formation if present at certain stages of development (Rangel-Aldao, Allende and Cayama 1985; Inselburg 1983).

In eukaryotes the major cellular receptors for cAMP are the regulatory subunits of certain types of protein kinase.   Not all protein kinases are sensitive to cAMP, however.   Protein kinases are responsible for the covalent modification (by phosphorylation) of proteins, particularly certain regulatory enzymes, causing a major change in their activity.   In *T. brucei*, a number of cAMP receptors have been found.   They are possibly protein kinases, but their properties are different from these enzymes in other eukaryotes (Rangel-Aldao and Opperdoes 1984).   In contrast, *T. cruzi* has a single cAMP receptor which is not associated with a protein kinase.   Its binding properties for cAMP analogues are different from those described for the regulatory subunits of mammalian protein kinases. The function of this cAMP receptor is unknown but it appears to be unique and a cAMP analogue may be found that will bind selectively to the *T. cruzi* receptor.

Protein kinases, whether cAMP-sensitive or not, are important regulators of development and other processes (e.g. energy metabolism) in parasites.   A number of compounds inhibit these enzymes in Protozoa.   Micromolar concentrations of suramin inhibit protein kinase I in *T. b. gambiense*; the trypanosomal enzyme is considerably more sensitive to suramin than the mammalian one (Walter 1980).   In

P. berghei, protein kinase activity increases with maturation of the intracellular parasites. Its activity is stimulated by spermine and spermidine but it is not sensitive to Ca^{2+} or cAMP. Quercetin is a potent inhibitor of this enzyme *in vitro* (Wiser, Eaton and Sheppard 1983).

In helminths, protein kinases, cAMP-dependent and -independent, have been identified in filariae, other nematodes, and schistosomes (see Ossikovski and Walter 1984). They are probably ubiquitous and presumably have a similar regulatory function in helminth metabolism as in mammals. In *A. suum* the activity of phosphofructokinase is determined in part by phosphorylation-dephosphorylation that is catalysed by a cAMP-dependent protein kinase (Hofer, Allen, Kaeini and Harris 1982). Pyruvate dehydrogenase from *A. suum*, on the other hand, is phosphorylated by a specific cAMP-independent protein kinase (Komuniecki, Wack and Coulson 1983). Suramin has been shown to inhibit protein kinase I (cAMP-independent) in *O. volvulus*, which may contribute to its filaricidal activity (Walter and Schulz-Key 1980). The cAMP-independent protein kinases of *A. galli* are inhibited by suramin *in vitro*, and one enzyme also by stibophen (Ossikovski and Walter 1984).

Calmodulin

Calmodulin is an intracellular Ca^{2+} receptor that plays an important part in modulating the activity of certain enzymes and cellular functions in eukaryotic cells. It is widely distributed in the protozoan, animal and plant kingdoms. Calmodulin with bound Ca^{2+} is a potent activator of many intracellular enzymes and processes. These include Ca^{2+}-dependent cAMP-phosphodiesterase and adenylate cyclase, phosphorylase kinase, Ca^{2+}-transporting ATPase, microtubule assembly, lipid and carbohydrate metabolism. Ca^{2+}-calmodulin also binds to the phenothiazine drugs *in vitro*, a property often used in its purification. The Ca^{2+}-calmodulin-phenothiazine complex is less effective than Ca^{2+}-calmodulin inactivating target enzymes such as cAMP-phosphodiesterase.

A protein identified as calmodulin has been isolated from *H. diminuta* (Branford White, Hipkiss and Peters 1984). Its properties closely resemble those of mammalian calmodulin and it is equally susceptible to inhibition by the phenothiazine compound, trifluoperazine. Parasites exposed to this compound *in vitro* showed a breakdown in cellular integrity that was Ca^{2+}-dependent, and a Ca^{2+}-dependent activator of cAMP phosphodiesterase (presumably calmodulin) was inhibited by trifluoperazine (Branford White and Hipkiss 1984). Whether this property is important in the anthelmintic activity of phenothiazine *in vivo* has not been demonstrated.

Protozoan parasites are sensitive to Ca^{2+} in their environment and also contain calmodulin. Calcium given to *T. brucei in vitro* causes changes in the activities of adenylate cyclase and endoribonuclease, and causes the release of variable surface glycoprotein (VSG) (see Ruben and Patton 1985). Calmodulin in African trypanosomes has different properties from the mammalian protein, and certain phenothiazine calmodulin antagonists are toxic to *T. brucei* and *L. donovani in vitro* (Ruben and Patton 1985). Trifluoperazine interferes with Ca^{2+} function in *T. b. rhodesiense* by binding to calmodulin (Ruben, Egwuagu and Patton 1983). *T. cruzi* epimastigotes are killed by chlorpromazine and fluphenazine, which have been shown to bind to calmodulin in the parasites and prevent activation of cAMP-phosphodiesterase (Téllez-Iñón, Ulloa, Torruella and Torres 1985). Clearly, there is chemotherapeutic potential in compounds such as these that selectively inactivate calmodulin in parasites.

Polyamines
The importance of polyamines (putrescine, spermidine, spermine) in regulating development and other processes in Protozoa is beginning to come to light. These

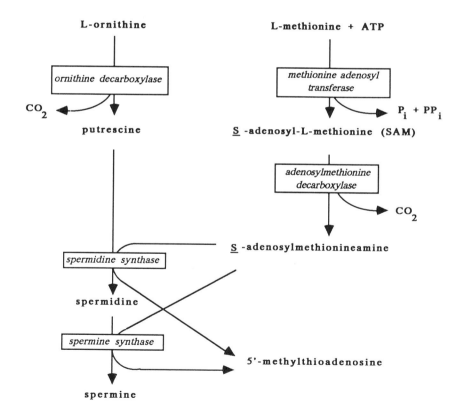

Figure 5.4: Pathways of Polyamine Biosynthesis

low molecular weight cationic compounds appear to have a multitude of biochemical functions. In particular, they bind ionically to nucleic acids and play a role in the cell cycle, in cell division and in differentiation (see Pegg and McCann 1982; Grillo 1985). Their exact sites of action in these processes have, in general, not yet been elucidated.

Polyamines have been investigated in parasitic Protozoa because of their apparent importance in the cell cycle. They are not readily taken up from the medium by Protozoa but are synthesised in the cell and their concentrations vary at different stages of the life/cell cycle. In trypanosomes they may be important in regulating or controlling replication of kinetoplast DNA; in *T. b. brucei*, spermidine and spermine have been shown to activate a DNA polymerase that has different properties from the mammalian enzyme (see Bacchi 1981). Polyamines have other roles too. Spermidine and spermine, for example, may act as coenzymes for NAD-linked glycerol-3-phosphate dehydrogenase in trypanosomes (see Bacchi 1981). In trypanosomes and leishmaniae the enzyme glutathione reductase requires a unique cofactor, trypanothione, that is a conjugate of glutathione and spermidine (Fairlamb, Blackburn, Ulrich, Chait and Cerami 1985). *L. m. mexicana* promastigotes and *T. vaginalis* excrete significant amounts of putrescine into the culture medium; the function of this is not understood (White, Hart and Sanderson 1983; Coombs and Sanderson 1985).

It appears that polyamines may be especially important to parasitic Protozoa because of their role in cell multiplication. Compounds that interfere with polyamine function or synthesis are therefore likely to be useful antiprotozoals. The effectiveness of the cationic trypanocidal compounds such as quinapyramine or ethidium may be due, at least in part, to competition with polyamines for intracellular binding sites (see Bacchi 1981), thus interfering with polyamine function. Compounds that inhibit synthesis of polyamines also have great potential for damaging parasite metabolism; two enzymes in the synthetic pathway (see Figure 5.4) have so far proved to be vulnerable:

(i) *Ornithine decarboxylase* (ODC), which is inducible, has a short half-life and is rate-limiting in the pathway depicted in Figure 5.4, is inhibited by the substrate analogue α-difluoromethylornithine (DFMO), which binds covalently to the enzyme (see McCann, Bacchi, Nathan and Sjoerdsma 1983). Selectivity is due to the fact that the trypanosome enzyme is more susceptible (*in vitro*) than that of the host, and that the parasites actively transport DFMO whereas it enters host cells only by diffusion. DFMO is effective *in vivo* against bloodstream trypanosomes, and also *Eimeria*; *in vitro* it also affects *P. falciparum* at certain stages, and *Giardia* (see Meshnick 1984a; McCann et al. 1983). In *P. berghei* exoerythrocytic schizogony is inhibited by DFMO, but not intraerythrocytic schizogony (Hollingdale, McCann and Sjoerdsma 1985). Chloroquine also inhibits ODC in *P. falciparum in vitro* (Königk and Putfarken 1985). In trypanosomes, inhibition of ODC is quite devastating: it prevents putrescine synthesis, causing rapid depletion of polyamines, it inhibits DNA and RNA synthesis, and blocks cell division, causing morphological and metabolic aberrations. S-Adenosylmethionineamine (decarboxylated S-adenosylmethionine), the co-substrate for spermidine synthase, accumulates; this may deplete the supply of adenine for nucleic acid synthesis (see Figure 5.5).

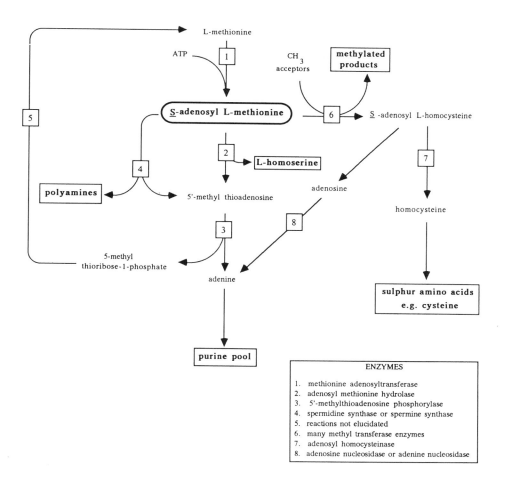

Figure 5.5: The Central Role of S-Adenosyl L-Methionine (SAM) in Cellular Metabolism (modified from Whaun, Brown and Chiang 1984)

Synergistic effects between DFMO and certain other trypanocidal compounds, particularly bleomycin, have been observed (see McCann *et al.* 1983). Such combinations are especially effective against infections of the central nervous system. The synergy may be due to depletion of polyamines by DFMO, exposing polyamine binding sites that are not normally free for the binding of other compounds. Bleomycin contains an amine moiety that binds to DNA; if trypanosome DNA becomes exposed as a result of polyamine depletion, the binding and effectiveness of bleomycin in damaging DNA would be enhanced.

(ii) *Adenosylmethionine decarboxylase* supplies the substrate *S*-adenosylmethioninamine for conversion of putrescine to spermidine. Inhibition of this enzyme thus prevents spermidine synthesis. Methylglyoxal *bis*(guanylhydrazone), an analogue of spermidine, inhibits this enzyme in both mammals and trypanosomes (see Meshnick 1984a). It kills bloodstream *T. b. brucei in vivo* but is not curative. Recently-developed analogues of this compound have proved more effective, and there is potential for further development in this area of research. In addition, it has been shown that berenil and pentamidine, which are structurally related, inhibit this enzyme from rats, both *in vitro* and *in vivo* (Karvonen, Kauppinen, Partanen and Pösö 1985).

Methylation reactions
S-Adenosylmethionine and its derivatives, 5'-methylthioadenosine and *S*-adenosylhomocysteine, play central roles in methyl group transfer reactions (e.g. methylation of DNA, proteins), in synthesis of sulphur-containing amino acids and in recovery of adenine for purine metabolism (see Figure 5.5), though their precise roles in parasites have not been described. Analogues of these compounds have shown some promise as antiprotozoals because they have potential to disrupt many different areas of essential metabolism in the parasites. These compounds include sinefungin, which is active against trypanosomes *in vitro* and *in vivo* (Meshnick 1984b), and also against *P. falciparum in vitro*, and *S*-isobutyladenosine and deazaadenosine, which may inhibit methyltransferase reactions in *P. falciparum in vitro* (Whaun, Brown and Chiang 1984; see also Desjardins and Trenholme 1984).

5.3.10 *Cellular integrity*

The overall structural integrity of cells, their external and internal membranes and organelles, and the physical and biochemical properties of their external surfaces are all essential for survival. It is crucial for parasites to have an intact external surface, since this is in constant contact with host-derived molecules and cells. These molecules or cells may be essential for the parasites' nutrition or development, or they may be potentially damaging to the parasites. Agents that interfere with parasite cellular integrity will either kill them directly, by causing lysis or by interfering with cell movement or intracellular organisation, or will damage the external surface so that important receptors or transport molecules are inactivated, or so that the surface is exposed to attack by the host's immune system. In this section we shall discuss important aspects of cellular integrity in parasites, and how these may be exploited by chemotherapy.

The external surface and receptors in parasitic Protozoa

At the molecular level, the topography and the chemical properties of the external surface of cells are determined by the glycoproteins, proteins, lipopolysaccharides and other molecules that are attached to or embedded in the cell membrane. Of great importance are the carbohydrate moieties of glycoproteins and other glycosylated molecules which form a characteristic, protective oligosaccharide matrix around the cell membrane. This protective layer is open at intervals to expose molecules and pores on the cell surface that are part of the cell membrane. These molecules have specific functions for, for example, recognition or transmembrane transport; their shapes and chemical properties permit interaction only with ligands with a specific steric configuration or specific chemical constitution. In addition, the overall topography of the external carbohydrate network limits the type of molecule (as defined by charge, size, shape, etc.) that can have access to the molecules on the membrane.

The cell membrane and its associated structures are constantly being renewed by processes of invagination of the old membrane (endocytosis) and fusion of vesicles of new membrane (exocytosis). In this way damage to the membrane is repaired, 'receptor' molecules are renewed or changed, secretion occurs, molecules or food particles are internalised, developmental changes can take place, cells can grow and move. The molecular processes controlling these events in cells are not well understood, but certain agents have been discovered that affect the integrity of the external surface in parasites.

The antibiotic tunicamycin inhibits the glycosylation of proteins by preventing the formation of asparagine-linked oligosaccharides. This compound is not selective and is not therefore used clinically, but it has been used to demonstrate the importance of glycoproteins in parasites. Treatment of *T. cruzi in vitro* causes loss of cell surface glycoproteins; these treated parasites are unable to infect mammalian cells because they are not taken up by the cells (Zingales, Katzin, Arruda and Colli 1985). Thus, parasite surface glycoproteins are essential for host cell 'recognition' of *T. cruzi*. *L. donovani* promastigotes treated with tunicamycin *in vitro* are not infective to macrophages either *in vitro* or *in vivo*; in this case the parasites are taken up by the host cells, but die in the phagolysosomes (Nolan and Farrell 1985). In this species, therefore, glycoproteins are not essential for internalisation by the host cell, but have a role in protecting the parasites from proteolytic attack in the host's phagolysosomes. The importance of the variable surface glycoprotein (VSG) in protecting bloodstream African trypanosomes from attack by the host's immune system is well established; tunicamycin pre-treatment of *T. brucei* improves the survival of mice after infection (see Meshnick 1984a).

Swainsonine, a mannosidase inhibitor, interferes with carbohydrate processing by preventing the hydrolysis of mannosides, causing accumulation of mannose-containing oligosaccharides in cells. *T. cruzi* treated *in vitro* with this alkaloid produce abnormal oligosaccharides with a high content of mannose; the parasites are less infective to host cells *in vitro* because host cell-parasite association is diminished (Villalta and Kierszenbaum 1985). The effect can be reversed by treating the parasites with mannosidase. Swainsonine has similar effects on host cells.

If agents such as these that act against surface glycoproteins can be found that show some selectivity for parasites, they will be powerful weapons because they have the potential to disrupt the life cycles of intracellular parasites which are otherwise difficult to treat. They may also expose the parasite surface to attack by the host's immune system, thus circumventing the problem of antigenic variation. Regrettably, there are few clues to selectivity available at present. A possibility in the case of *T. brucei* is ethanolamine, which is entirely derived from the host and is essential for VSG production by these parasites because it forms part of the lipopolysaccharide anchoring the VSG to the plasma membrane. A compound interfering with either transport of ethanolamine by the parasite, or its incorporation, may well be sufficiently selective to be useful (Rifkin and Fairlamb 1985).

Evidence is accumulating that specific 'receptor' molecules on host and parasite cells are responsible for cell-cell recognition. Inactivation or removal of such receptors by chemical or immunochemical means would interrupt the infection process in many parasites. There are several groups of compounds known to affect cell surface receptors. For example, certain lysosomotropic agents, which include primaquine, chloroquine, NH_4Cl, and methylamine, are weak bases that accumulate in acidic intracellular compartments, especially lysosomes, and disrupt their functions. One important function that is affected is the recycling of membrane receptors from endocytotic vesicles. Primaquine and chloroquine inhibit the binding and entry of *P. berghei* sporozoites to hepatoma cells *in vitro*, an effect probably due to depletion of the membrane receptors on the host cells, which would result from inhibition of receptor recycling (Schwartz and Hollingdale 1985). There is no direct effect on the sporozoites. This particular effect of primaquine or chloroquine is not considered responsible for their antimalarial activity, but it shows that cell surface receptors can be depleted, thus interrupting the infection process when cell-cell interaction or other receptor-mediated activity is required. In this particular example the effect was on host cells; agents having selective effects on parasite cells would clearly be particularly valuable.

Compounds that affect the permeability of the cell membrane, or specific trans-membrane transport processes in parasites (or their host cells, in the case of intracellular Protozoa) are potential chemotherapeutic agents. The nutritional requirements of parasites are different from their hosts, because of their limited biosynthetic capabilities and reliance on salvage mechanisms. Therefore, specific transport molecules or systems may be present in the external membranes of parasites that are not represented in the host. This has been discussed in Chapter 3 but we shall present some slightly different examples here.

L. donovani and *L. major* promastigotes and amastigotes possess a H^+-ATPase on their external membranes that maintains a proton electrochemical gradient across the membrane. This gradient drives the active transport of nutrients such as proline and glucose. In host cells, this ATPase is found only in the mitochondrial membranes. The tricyclic antidepressants clomipramine and nitroimipramine kill both extracellular promastigotes and intracellular amastigotes without damaging the host macrophages (Zilberstein and Dwyer 1985). Proline transport by promastigotes was inhibited by clomipramine and it appears that these compounds disrupt the proton gradient at the cell membrane, either by directly affecting the ATPase, or by ionophore activity.

The ionophore monensin stimulates the (Na^+-K^+)-ATPase in the external membrane of *Eimeria tenella* sporozoites *in vitro* by increasing the influx of Na^+ ions through the membrane. This leads to swelling and death of the parasites. A related compound, narasin, causes swelling of intracellular sporozoites without affecting the host cells. This suggests a difference between host and parasite cell membranes that affects the binding of narasin (Smith and Strout 1980).

It is clear that more potential drug targets will be revealed that fall into this category, and, as discussed above, they will probably be especially useful in the chemotherapy of intracellular parasites. It should be noted in this context that host cells containing intracellular parasites may have altered permeability properties and altered surface molecules or 'receptors'. Such differences between infected and uninfected cells may also prove useful in chemotherapy.

Tegumental integrity in helminths

The tegument in helminths functions not only as a surface, protecting the parasites from potentially hostile interaction with the host, but also as an absorptive, secretory and highly metabolically active interface, especially in cestodes and Acanthocephala where it includes the functions of the gut. It also has an important role in controlling water and ion uptake in helminths. Tegumental membranes, being cell membranes, are constantly being resorbed, recycled and renewed. Therefore, functional and physical integrity of the tegument is essential for a healthy parasite and disruption would be, in most cases, lethal. For helminths living in body tissues in close association with host cells, a break or lesion in the tegument may not necessarily be directly lethal, but would expose the parasite to attack by the host's immune system, which would ultimately kill the parasites. A number of currently-used anthelmintic compounds affect the structural integrity of the tegument but, in most cases, the exact site of attack has not been identified and we do not know whether the damage is direct, or a result of, say, irreversible inhibition of a vital biosynthetic process or uptake of an essential nutrient. It appears that trematodes and cestodes are more susceptible to tegumental damage than the cuticle-bearing nematodes or Acanthocephala.

Diamfenetide, a flukicide, interferes with tegumental ion regulation in *F. hepatica*, causing depolarisation of the tegument and swelling of the parasite. *In vitro*, this occurs within 30 minutes after exposure, which suggests that it is most likely to be the primary effect of this compound, though its precise molecular site of action is unidentified and comparisons with *in vivo* treatment have not yet been made (Rew and Fetterer 1984).

Three antischistosomal compounds also cause damage to the tegument. Schistosomes have a double tegumental membrane, the outer layer of which has properties different from those of a normal plasma membrane. It has a rapid turnover and its major function appears to be immunological protection. Hycanthone or oxamniquine treatment *in vivo* both cause tegumental damage, but the effect is delayed until 4-10 days after treatment (see Coles 1984), which suggests an indirect attack. Interestingly, both compounds affect male schistosomes to a greater extent than females. The host's immune system is ultimately responsible for eliminating the parasites which have lost their immunological protection.

Praziquantel, on the other hand, has a rapid and dramatic effect on schistosomes. Within 30 seconds of exposure to 1 mM praziquantel *in vitro*, the tegumental membranes of *S. mansoni* become vacuolised and, at the same time, the muscles rapidly contract and remain paralysed (Bricker, Depenbusch, Bennett and Thompson 1983). These effects are partially antagonised by high concentrations of Mg^{2+}. Ca^{2+} is required in the medium to achieve muscular paralysis, but the tegumental damage is only partially inhibited by lack of externally-supplied Ca^{2+}. There is a rapid influx of Ca^{2+} into the worms following treatment; this influx may have a direct effect on the cytoskeleton and on muscle contraction, but Ca^{2+} does not affect transport of the drug. Praziquantel does not act as an ionophore, and does not affect ATPase activity (see Andrews, Thomas, Pohlke and Seubert 1983). Similar damage to schistosomes is caused by the calcium ionophore A-23187 and, interestingly, the benzodiazepine Ro 11-3128. Praziquantel also damages the tegument of the anterior parts of cestodes and causes paralysis, but in this case there is a rapid *efflux* of Ca^{2+} from the worms (Prichard, Bachmann, Hutchinson and Köhler 1982). The difference in direction of movement of Ca^{2+} in cestodes and schistosomes may reflect different mechanisms of calcium homeostasis in the two parasite groups: cestodes probably use stored Ca^{2+} for muscle contraction, whereas the schistosomes remove it from the surrounding medium when required.

It is not clear whether praziquantel-induced muscular paralysis and tegumental damage in these parasites are causally related, or whether these two processes have in common a Ca^{2+}-dependent process that is disrupted by praziquantel. Ca^{2+} is intimately involved in both muscle contraction and many aspects of cell movement controlled by the cytoskeleton. When these are disrupted by high concentrations of Ca^{2+}, the result would be paralysis, blebbing and the prevention of normal organelle and vesicle movement within the cells. The vacuolisation starts at the base of the syncytial layer and the tegument is effectively disrupted from within, possibly as a result of interference with lysosome function (Leitch and Probert 1984). Elucidation of the precise site(s) of action of this compound will identify an important target for chemotherapy in these parasites and will also permit some insights into the processes controlling the maintenance of tegumental integrity.

Schistosomes possess a low-affinity benzodiazepine binding site that appears to be located on the tegument and that has different properties from the high-affinity benzodiazepine binding sites found in vertebrates (Bennett 1980). This is of interest because certain benzodiazepines, e.g. Ro 11-3128, are schistosomicidal (but not against all the medically-important schistosome species) and appear to act in a similar way to praziquantel, by causing an increased influx of Ca^{2+} accompanied by tonic muscular paralysis and disruption of the tegument (see Bennett and Depenbusch 1984). Praziquantel is not a benzodiazepine, and its binding site is different from that of benzodiazepines. Therefore there appear to be at least two apparently unique sites, of considerable chemotherapeutic importance, on the schistosome tegument. These sites are related by their control of Ca^{2+}-dependent events that cause muscle contraction and disrupt tegumental integrity.

Membrane integrity
The chemical composition of parasite membranes differs from that of host membranes, which means that parasite membranes may bind and transport drugs differently. In addition, some protozoan groups have unique intracellular organelles, the glycosomes and hydrogenosomes, whose membrane composition might be unusual and susceptible to selective membrane-acting compounds. These agents may cause direct lysis of membranes, or cause them to become 'leaky', or interfere with their digestion or recycling in lysosomes, or change their physical properties such that they no longer function normally. In this section we shall discuss some examples of compounds that specifically affect parasite membranes.

Dihydroartemisinine, an experimental antimalarial compound derived from artemisinine (qinghaosu), is accumulated by malaria parasites *in vitro*, but not by uninfected erythrocytes. In *P. falciparum* treated *in vitro*, it is localised in the cell membrane of the parasites and in the membranes of the digestive vacuoles. In *P. berghei*, swelling of the mitochondria and endoplasmic reticulum and damage to the cell membrane occur within 30 minutes after treatment *in vivo*. Such early effects suggest direct membrane damage or ionophore activity, leading to permeability changes and swelling of the cells (Ellis, Li, Gu, Peters, Robinson, Tovey and Warhurst 1985). Subsequent effects of treatment include inhibition of protein and, later, nucleic acid synthesis. The structure of dihydroartemisinine, a sesquiterpene, is such that it could enter a membrane of suitable composition and alter its permeability, or, perhaps, have a direct effect on particular components of the membrane.

Mefloquine potentiates the effect of artemisinine against *P. berghei in vivo*; this compound binds to phospholipids in membranes and has a higher affinity for membranes of infected erythrocytes than uninfected erythrocytes, which suggests that the membranes of infected cells have a higher lipid content (see Howells 1985). Chloroquine also interacts with membranes and partially inhibits uptake of dihydroartemisinine.

Chloroquine binds with high affinity to ferriprotoporphyrin IX (FP), also called haemin or haematin, a toxic product of haemoglobin digestion that lyses erythrocytes. The chloroquine-FP complex also lyses cells, by membrane damage that first appears as loss of ability to maintain cation gradients, causing swelling. FP is released by some malaria parasites digesting haemoglobin; it is membrane-bound and is normally inactivated by precipitation as malaria pigment (haemozoin) (see Fitch, Dutta, Kanjananggulpan and Chevli 1984). Other antimalarial compounds – amodiaquine, mepacrine, mefloquine and quinine – also bind to FP to form membrane-soluble complexes, but with lower affinity. Chloroquine-resistant strains of *P. berghei* do not accumulate haemozoin and appear to digest haemoglobin by a route that does not release FP (Wood, Rock and Eaton 1984). Thus, one important effect of these antimalarial compounds is membrane damage brought about by drug-mediated interruption of the detoxification of FP.

Intracellular coordination
The movement of organelles, of cells, cell division, feeding and digestion, renewal of membranes, secretion and many other cellular processes require coordination and proper functioning of the organelles and the molecules that effect their movement within the cell. Rapidly feeding and growing parasites are more susceptible to agents disrupting these processes than less active host cells; such processes warrant further investigation as potential targets for chemotherapy.

Lysosomes are essential organelles responsible for a cell's digestive processes. Disruption of their function is ultimately lethal. Some 'lysosomotropic' agents are weak bases that inhibit intralysosomal digestion of membrane constituents. Chloroquine, for example, interferes with the digestion of endocytotic vesicles in the food vacuole of *P. falciparum in vitro* (Yayon, Timberg, Friedman and Ginsburg 1984). The host-derived membranes of these vesicles are not digested and accumulate in an enlarged food vacuole, where the chloroquine and the parasite's pigment deposits also accumulate. This may be due to altered pH in the food vacuole, or to a direct effect of chloroquine on the digestive enzymes. A number of anthelmintic agents may affect lysosomal function (see Leitch and Probert 1984) but this has not been unequivocally demonstrated.

Other agents also affect lysosomes. A non-ionic detergent macromolecule, Triton WR-1339, which cannot be digested in lysosomes, has been successfully used against *T. brucei in vitro* and *in vivo* (Opperdoes and Van Roy 1983). It is taken up by cells during feeding and accumulates in the phagolysosomes, causing the development of a massive lysosomal vacuole with leaky, fragile membranes. Death of the trypanosomes would be due either to leakage of digestive enzymes into the cell, or to the impairment of normal lysosome function. This agent is also taken up by some host cells (e.g. macrophages). Whether it is possible to create compounds such as these that will be selective for parasites has yet to be demonstrated.

Microtubules, which are assembled in cells by the rapid polymerisation and depolymerisation of tubulin, are responsible for much of the intracellular movement of organelles and vesicles – thus participating in such processes as endocytosis, secretion, digestion, membrane maintenance and coating, – and for some aspects of cell movement, including cell division. Inhibition of tubulin synthesis, its polymerisation or depolymerisation, has devastating effects on parasites. Tubulin from parasites often appears to have slightly different binding properties from that of their hosts, and therein may lie the basis of the selectivity of those compounds that affect microtubules. The neuroleptic phenothiazine compounds are known to inhibit microtubule polymerisation in Protozoa; they have been shown to kill *T. brucei in vitro*, and *L. donovani in vitro* within their host macrophages, which are unaffected (Seebeck and Gehr 1983; Pearson, Manian, Harcus, Hall and Hewlett 1982). The effect is rapid, occurs at micromolar concentrations, and results in loss of motility, loss of pellicular microtubules, and degeneration of nuclear and cytoplasmic structure.

The substituted benzimidazole compounds, which are used as broad-spectrum anthelmintics, appear to inhibit tubulin polymerisation in susceptible helminths (see Behm and Bryant 1985; Lacey 1985). Ultrastructural studies after treatment *in vitro* of nematodes and cestodes have shown that an early effect of treatment (evident within 3-6 hours) is the disappearance of cytoplasmic microtubules, followed by degeneration of the affected cells (usually the tegumental or gut epithelial cells). Other functions affected by these compounds in some (but not all) helminths include inhibition of glucose and Na^+ uptake, inhibition of acetylcholinesterase secretion by some intestinal nematodes, inhibition of fumarate reductase activity, uncoupling in mitochondria, and embryotoxicity. Most of these effects could be secondary to inhibition of tubulin polymerisation but unequivocal evidence is lacking, especially in non-nematode parasites.

The binding properties of nematode tubulin for colchicine and the substituted benzimidazoles are different from those of mammalian tubulin, possibly a result of slight differences in the structure of the α-subunit (Dawson, Gutteridge and Gull 1983). However, although nematode tubulins are in general more susceptible *in vitro* to inhibition of polymerisation by these compounds, there is no absolute correlation of susceptibility with their *in vivo* activity. It is possible, therefore, that the selectivity of these compounds is due to the combination of a more susceptible molecular target (or targets) plus, possibly, differential uptake of these compounds and/or a reduced capacity of the parasites to detoxify them. Helminths have limited abilities to detoxify organic compounds (Munir and Barrett 1985).

Some of these compounds also affect microtubule polymerisation in the host: this has been exploited by Hennessy (1985) and co-workers, who showed that inclusion of a low concentration of parbendazole, a potent inhibitor of microtubule assembly, in a fenbendazole sheep treatment, reduced the host's biliary secretion of the active metabolites of fenbendazole, thus increasing the plasma concentrations of these metabolites and effectively potentiating the treatment.

Susceptibility to oxidation
The cells of organisms can be severely damaged by free radicals, which are highly reactive, unstable and donate their unpaired electrons readily to biological molecules. Free radicals derived from oxygen (singlet oxygen $^1O_2$, superoxide anions O_2^- or hydrogen peroxide (hydroxyl radicals $OH\cdot$) are very common; other radicals may arise from chlorine compounds, for example, or from certain free radical generating compounds, or they may be generated during normal metabolism or during cellular detoxification reactions (Figure 5.6; for a complete discussion of

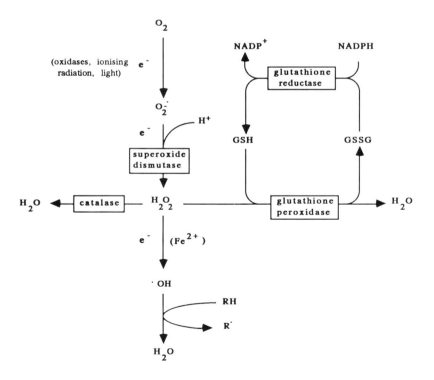

Abbreviations: GSH, reduced glutathione; GSSG, oxidised glutathione; RH, any molecule with a removable proton (lipid, protein, nucleic acid, etc.)

Figure 5.6: Intracellular Single Electron Reduction Pathways of Oxygen

this subject see, for example, Docampo and Moreno 1984; Halliwell and Gutteridge 1985; and Clark, Hunt and Cowden 1986). The most damaging of the oxygen-derived free radicals is the hydroxyl radical. Especially vulnerable to oxidation by free radicals are cell membranes, because their lipids are susceptible to peroxidation; this is a self-sustaining process (see Figure 5.7). DNA and proteins are also readily damaged by free radicals. Oxygen-derived free radicals are found at low concentrations in the environment wherever oxygen is present, but their levels are elevated in those cells where oxygen participates in metabolism.

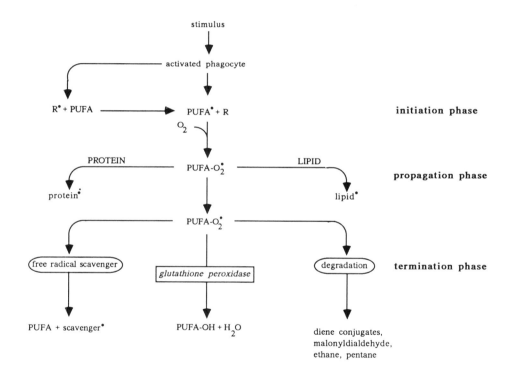

PUFA: polyunsaturated fatty acid; $R^\bullet$: free radical

Free radical scavengers include reduced glutathione, other thiols (for example, protein thiols), vitamin E, ascorbic acid and β-carotenes. Scavenging enzymes include superoxide dismutase, catalase and glutathione reductase.

Figure 5.7: The Sequence of Events in Free-Radical Mediated Lipid Peroxidation (after Fantone and Ward 1982).

The complete reduction of oxygen to water requires four electrons per molecule of oxygen. If these electrons are transferred one at a time, the intermediate species are, in sequence, superoxide anion, hydrogen peroxide, hydroxyl radical (Figure 5.6). The only enzymes that transfer all four electrons simultaneously are the copper-containing oxidases, of which the most important is cytochrome oxidase. Oxidases containing other transition metals, usually Fe – these include the flavoproteins, ferredoxins and the majority of cytochromes – generate superoxide anions, which leak continuously from the electron transport chains of the mitochondria and endoplasmic reticulum. Therefore, organisms that reduce oxygen – this includes the majority of parasites – will generate superoxide and its derivatives within their cells. The reactions of the aerobic electron transport chains are the most important source of superoxide in most organisms.

Another especially important source of superoxide in host-parasite systems is the host's membrane-bound NADPH oxidase, a flavoprotein present in white blood cells, especially inflammatory and phagocytic cells (see Badwey and Karnovsky 1980). This enzyme reduces oxygen to superoxide, which is released into phagocytic vacuoles and also into the surrounding medium during the 'respiratory burst'. Superoxide radicals, and radicals derived from them, are responsible for killing invading microorganisms, but they are also responsible, in some cases, for pathological damage to the host (see Clark, Cowden and Hunt 1985) because their release may damage surrounding host tissues.

Living organisms are protected from the damaging effects of free radicals by protective enzymes, which metabolise the radicals to harmless products, or by certain free radical scavenging compounds, termed antioxidants. Some of these are listed in Figure 5.7. If a cell's capacity to detoxify free radicals is exceeded, membrane damage, which ultimately leads to cell lysis, will result. Free radical damage to nucleic acids and proteins will also interfere with the cell's ability to grow and reproduce. Parasites living inside their hosts are exposed to oxygen-derived free radicals not only from their own metabolism but also from oxygen present in the host's tissues, or that released as superoxide from activated white blood cells in the vicinity. Intracellular parasites that reside within red blood cells are normally exposed to considerable oxidant stress because of both the high concentrations of oxygen in these cells and the autoxidation of haemoglobin which is a radical generating process. They therefore require efficient oxidant protection at all times. Similarly, because of the secretion of superoxide into phagolysosomes, parasites living in macrophages may be exposed to significantly higher levels of oxidants than parasites inhabiting other types of cell.

Many parasites live under conditions where the oxygen tension is low, in the gut or the urinogenital tracts, for example. Their levels of oxidant *protection* may therefore be low, which may be exploited in chemotherapy since most host cells are well protected against oxygen-derived free radicals. Levels of protection against oxidants may vary at different stages of the life cycle. Some parasite groups lack certain protective enzymes altogether (see, for example, Smith and Bryant 1986).

In addition, agents which increase the inflammatory reaction may also be useful, provided it is appropriately located and can be specifically directed against the parasites. Therefore, several strategies could be used in designing chemotherapy that will exploit the sensitivity of parasites to oxidation:

(i) Search for particular deficiencies or differences in their protective enzymes or antioxidant content, then stress them with oxidants designed to take advantage of these, or activate the host's cellular immune system to attack the oxidant-sensitive parasites in specific ways.

(ii) Search for unique reducing (or 'detoxification') pathways in the parasites and present them with compounds that form free radicals when they become reduced.

(iii) Search for specific uptake mechanisms in parasites that will concentrate free radical generators that dissolve on or bind specifically to parasite membranes.

Some compounds in current or experimental use utilise some of these strategies; we shall discuss some selected examples briefly here.

African trypanosomes lack the protective enzymes that contain haem, such as catalase, and are susceptible to H_2O_2 *in vitro* (Penketh and Klein 1986). It has been shown that agents that split hydrogen peroxide (e.g. haem) release radicals that cause lysis of the parasites (see Howells 1985). Analogues of haem that do not bind to serum proteins are especially effective, and their effect is potentiated, both *in vivo* and *in vitro*, by arsenical compounds (e.g. melarsen) that deplete GSH in the parasites, thus increasing their sensitivity to oxidants.

Glutathione in its reduced form plays such a central role in protection against oxidants in most organisms that inhibition of glutathione synthesis or recycling is potentially lethal. Depletion of GSH is dangerous for organisms. At the same time, accumulation of oxidised glutathione (GSSG) is harmful to cells because it forms mixed disulphides with proteins and other molecules containing thiol groups. Important enzymes known to be inhibited by GSSG accumulation include adenylate cyclase, 6-phosphofructokinase and phosphorylase phosphatase (see Halliwell and Gutteridge 1985).

Organic arsenical compounds inhibit glutathione reductase from *L. carinii* adults, both *in vitro* and *in vivo*; the parasite enzyme is more susceptible to inhibition that the mammalian one (Bhargava, Le Trang, Cerami and Eaton 1983). Specific inhibitors of glutathione reductase have been shown to inhibit growth of *P. falciparum in vitro* and to cure mice infected with *P. vinckei* (Schirmer, Lederbogan, Eisenbrand and Königk 1984). The antischistosomal compound oltipraz causes depletion of GSH in *S. mansoni in vivo*, within 48 hours of administration. At the same time, GSH levels in the host tissues rise, which suggests an important difference in GSH metabolism between host and parasite that is worth investigating (Bennett and Depenbusch 1984). Oltipraz is potentiated by coadministration of cysteine, which may also inhibit GSH synthesis in the worms (Bueding, Dolan and Leroy 1982).

Another enzyme of glutathione metabolism that is susceptible to attack is glutamate-cysteine ligase, the first enzyme in the pathway of glutathione synthesis. *T. brucei* infections in mice have been cured by treatment with buthionine sulphoximine, which inhibits this enzyme (Arrick, Griffith and Cerami 1981). It is clear, therefore, that all aspects of glutathione metabolism and synthesis could be suitable targets for selective chemotherapy in some parasites.

E. histolytica is unique amongst eukaryotes because, when grown axenically, it does not possess the enzymes of glutathione synthesis or metabolism (Fahey, Newton, Arrick, Overdank-Bogart and Aley 1984). When these organisms are grown in the presence of bacteria, however, their resistance to oxidation increases, suggesting that they derive some protection (enzymes or scavenging compounds) from ingested bacteria (Bracha and Mikelman 1984). There may be some chemotherapeutic potential in this relationship because the parasite may have limited capacity to protect itself against oxidant attack, artificial or natural.

Superoxide dismutase in trypanosomatids resembles the prokaryotic enzyme in that it contains Fe instead of Cu/Zn or Mn that is found in the eukaryotic enzymes. This should render the enzyme susceptible to selective inhibition by compounds (for example iron chelators) that bind to the Fe-containing enzyme only (see Meshnick 1984b). This enzyme in *T. foetus* also contains Fe and may be susceptible in the same way (Kitchener, Meshnick, Fairfield and Wang 1984).

The electron transport systems of parasitic Protozoa have not been completely characterised in any of the taxonomic groups, but the differences in pathways and redox potentials of the components are the bases for the activity of some effective antiprotozoal drugs. For example, metronidazole, which kills anaerobic organisms only, is reduced, in the absence of oxygen, at the nitro group to produce a series of highly reactive, unstable radical intermediates that damage the parasite's hydrogenosomal enzymes, its DNA and other intracellular molecules and systems (see Müller 1986). This reduction occurs, in trichomonads, via that part of the hydrogenosomal ferredoxin-linked electron transport pathway that normally catalyses the reduction of protons to form hydrogen. Electrons are supplied from the parasites' energy metabolism via pyruvate:ferredoxin oxidoreductase. Metronidazole competes with protons for the electrons and is reduced to the nitro anion radical, which subsequently gives rise to more reactive and damaging radicals.

$$R-NO_2 + e^- \rightarrow R-NO_2 \rightarrow \text{toxic radical products}$$

When oxygen is present, however, the nitro anion radical, once formed, directly reduces oxygen to superoxide (which is then removed), regenerating metronidazole and thus breaking the chain and preventing formation of the more toxic metabolites (Moreno *et al.* 1984).

$$R-NO_2 + O_2 \rightarrow R-NO_2 + O_2$$

In addition, oxygen competes directly with metronidazole for electrons from the electron-donating pathway and diminishes the rate of metronidazole reduction. In aerobic organisms the nitro group is not reduced and the compound is not active.

Nifurtimox is also reduced to a nitro anion radical by the electron transport system in some parasites, but it is effective aerobically because the radical can be formed in the presence of oxygen; it then reduces oxygen to form a series of free radicals that inhibit growth. This compound is effective against *T. cruzi* because the parasite's defences against oxygen-derived free radicals are inadequate; it appears to attack the parasite's DNA (Goijman *et al.* 1985).

The naphthoquinones (e.g. menoctone, menadione, β-lapachone) are a group of compounds that resemble ubiquinone and appear to act by substituting for ubiquinone in trypanosomatid electron transport systems. They transfer electrons directly to oxygen, producing superoxide and hydrogen peroxide at concentrations that overwhelm the parasites' defences and cause lysis (see Docampo and Moreno 1984). Unfortunately these compounds are not effective against trypanosomes *in vivo* because they are apparently inactivated by the host.

A number of dyes and redox compounds, for example gentian violet, rose bengal or phenazine methosulphate, have shown activity *in vitro* against *T. cruzi* and *Leishmania*, including the intracellular stages (see Howells 1985). They appear to act by accepting electrons, either by enzymatic reduction within the parasites or via activation by light, forming free radicals that damage the cells, especially in the presence of oxygen.

Most helminths appear to possess the protective enzymes superoxide dismutase and various peroxidases, but have low or undetectable activities of catalase (see Docampo and Moreno 1984). In addition, the activities of these enzymes may vary at different stages of the life cycle, which means that they may be most susceptible to oxidants at specific phases of their life cycles. Eosinophils, which considerably increase in number during helminth infections, and also neutrophils and basophils, secrete large quantities of superoxide when stimulated. This has been shown to damage and kill *S. mansoni* schistosomula and *T. spiralis* new-hatched larvae *in vitro*. Helminths are therefore susceptible to oxidant attack, but there are few examples of anthelmintics in current use that have been shown to act in this way.

The schistosomicide niridazole is reduced at the nitro group by *S. mansoni* under anaerobic conditions, by a process that is NAD(P)H-dependent (Tracy, Catto and Webster 1983). Mammals do not reduce this compound. Exposure of schistosomes to niridazole *in vitro* causes depletion of non-protein thiols and covalent binding of reduced niridazole to the parasite's macromolecules, particularly proteins. Since thiols inhibit the covalent binding, but not the nitroreduction, it appears that reduced niridazole binds to the free sulphhydryl groups of enzymes and other macromolecules and inactivates them.

Carbon tetrachloride, an old remedy for liver fluke infections, probably acts via a free radical mechanism. This compound is metabolised in the host's liver to active radicals that cause lipid peroxidation, especially in membranes (see Clark *et al.* 1985). These radicals and their products cause considerable damage to the host's liver (which can regenerate), but cause greater damage to adult liver flukes because they are exposed to active metabolites concentrated in the bile. Hexachloroethane and hexachlorophene, also flukicides, may also act in a similar way.

5.3.11 *Neurophysiology and behaviour*

Parasites living freely in the blood, in the lymphatic system, in body cavities, or in the alimentary canal of their hosts need to be able to maintain their positions in the face of the movement of the host's tissues (e.g. peristalsis) or the flow of blood, lymph or digesta. This is especially true for the helminths which, being larger than Protozoa, are more susceptible to displacement by their host's activities. It is important, therefore, that they are motile and able to respond to changes in position. Motility of the whole animal or of specific organs is also important for other activities of helminths, such as feeding, migration, and reproduction. Some helminths possess holdfast structures such as hooks and suckers to maintain their positions. Many nematodes secrete an acetylcholinesterase into their environment, one of the roles of which may be to reduce the host's muscular activity in their vicinity.

Sensory reception is also very important in the lives of helminths – both free-living and parasitic – because it permits them to locate their niches, to respond to environmental stimuli, to move along a gradient, to migrate, to locate feeding sites, to find or recognise a mate, and possibly to repel other worms of the same sex. Many of the cues to which the parasitic or infective stages of parasites respond are provided by the host and may be beyond the realm of interference by chemotherapy, unless receptors with unique properties are identified. But, parasites also secrete hormones or pheromones that enable them to coordinate their various activities. For example, male *N. brasiliensis* respond to pheromones – apparently small peptides – that are secreted by females when they are receptive to mating, and there is also some evidence for production by males of a substance that repels other males (Bone 1982). By mechanisms such as these, which rely on chemoreception, worms can communicate with each other even at extremely low population densities. If we are able to identify the essential compounds involved in interparasite coordination, and also characterise the receptors for them, we may have some unique targets for chemotherapy. These molecular sites would probably be quite specific for a given species of helminth, and their blocking could be effective in disrupting the reproductive process.

A large number of anthelmintic compounds affect the motility of parasites (see Fairweather, Holmes and Threadgold 1984), either directly by acting on their nervous transmission or neuromuscular sytems, or indirectly by affecting ion transport or energy metabolism for example. It is not necessary for an immobilising agent to kill parasites: in most cases immobilisation for a certain length of time will cause them to be moved too far from their normal site in the host to regain their positions once they recover their motility. In addition, it may be appropriate in treating certain parasitic infections (e.g. schistosomes, some filariae) merely to immobilise the parasite's reproductive tract, for example, so that eggs or larvae are no longer released. Agents that directly paralyse or immobilise helminths are amongst the most effective and selective treatments available at the present time, especially for nematodes. In this section we shall discuss the actions of neuroactive anthelmintic drugs.

The nervous systems of helminths and other invertebrates differ considerably from those of vertebrates. In addition to basic differences in anatomy, the nerves are structurally different in that their axons are unmyelinated and are therefore more exposed to chemical attack directed against their membranes. In nematodes the motor nerves do not extend into the muscles; instead, muscle cells extend processes to the longitudinal nerve fibres, where synapses are formed. Nematode muscles are obliquely striated, in contrast to vertebrate muscles; cestode and trematode muscles, on the other hand, are similar in appearance to mammalian smooth muscle. The resting potential of nematode and trematode muscles ranges from -20 to -60 mV whereas in vertebrate muscles it is -90 mV. The neuropharmacology of helminths has not been investigated in detail for any species and it may be misleading to assume that classical neuropharmacological agents used in vertebrates act in the same way in helminths. Most pharmacological work in helminths has been performed on nematodes, especially *A. suum* because of its large size.

Table 5.1: Some Neurotransmitters Identified in Adult Helminths

	Excitatory	Inhibitory
S. mansoni	5-hydroxytryptamine	acetylcholine dopamine norepinephrine
F. hepatica	dopamine	acetylcholine noradrenaline
H. diminuta	5-hydroxytryptamine	acetylcholine
A. suum	acetylcholine	γ-aminobutyrate

The neurotransmitters of helminths are distributed differently within the nervous system and they may also have different functions from those of their hosts.   Different taxonomic groups of helminths employ different neurotransmitters, some of which are listed in Table 5.1. These neurotransmitters appear to have different functions in different groups. For example, in *A. suum,* acetylcholine is excitatory whereas in S. *mansoni, F. hepatica* and *H. diminuta* , it is inhibitory. Therefore, agents acting as antagonists or agonists of acetylcholine may have different effects in different helminth groups. There is evidence that the acetylcholine receptors in trematodes have different pharmacological properties from mammalian receptors (Mellin, Busch, Wang and Kath 1983). This is shown by the reduced effects of d-tubocurarine and atropine, classical antagonists of the mammalian nicotinic and muscarinic acetylcholine receptors respectively (see Mansour 1984). Nematodes, like arthropods, use γ-aminobutyrate (GABA) as an inhibitory neurotransmitter. GABA is also present in the platyhelminth groups, but it does not appear to function as a transmitter. In contrast to mammals, GABA-ergic nerves in nematodes (as in arthropods) are found outside the central nervous sytem, which renders them more susceptible to selective attack.

Many neurotoxic agents act by interfering with transmission at nerve-nerve synapses or at neuromuscular junctions. Interference is achieved in many ways, including:

(i) by binding to receptors on the post-synaptic membrane and altering the activity of the post-synaptic nerve or muscle. The binding may be of high or low affinity and non-competitive or competitive with the local neurotransmitter. The effects of the neurotoxic agent may be *antagonistic* – preventing the normal activity of the nerve or muscle – or *agonistic* – intensifying or potentiating the normal activity. The duration of these effects depends on the relative affinity of the receptor(s) for the binding agents and on whether the parasite (or host) is able to inactivate the interfering agent enzymatically.

(ii) by inhibiting the enzyme responsible for inactivating the neurotransmitter at the synapse (e.g. acetylcholinesterase), thus potentiating the action of the neurotransmitter.

(iii) by inhibiting the enzyme(s) responsible for synthesising the neurotransmitter in the presynaptic nerve (e.g. choline acetyltransferase), or inhibiting uptake of precursors (e.g. choline) for synthesis of the neurotransmitter, thus preventing signal propagation across the synapse.

Neurotoxic agents may also act directly on nerve membranes by altering their ion channels, thus destroying the normal polarisation of the membranes and inactivating the nerves.

Acetylcholine is a neurotransmitter in all helminths that have been investigated though, as noted above, it has an excitatory role in nematodes but is inhibitory in other parasite groups. Cholinergic nerves in nematodes are found both in the main nerve cords and at neuromuscular junctions; the acetylcholine receptors have properties similar to the vertebrate nicotinic ganglionic receptors (see Lewis, Wu, Levine and Berg 1980). Agonists of acetylcholine that are effective against nematodes include levamisole, pyrantel, bephenium and methyridine. These compounds have a structural resemblance to acetylcholine or nicotine and bind to the acetylcholine receptors in the nerve cords of the parasites, where they initiate depolarisation. Since the drugs are not inactivated by acetylcholinesterase, they cause sustained muscle contraction, i.e. spastic paralysis, in the worms. The basis for their selectivity is that the parasite's receptors have a greater affinity for the compounds than the host's, but in high concentrations the drugs affect the host too (see Aubry, Cowell, Davey and Shevde 1980; Hsu 1980).

Levamisole appears to act in a slightly different way from the other compounds, because it has a biphasic effect. *A. suum* or *N. brasiliensis* exposed to certain concentrations of levamisole for long periods of time *in vitro* can recover some movement, for reasons that have not been elucidated in detail. As discussed previously, levamisole also stimulates glycogen synthesis in nematodes; this could cause a depletion in available ATP for muscle contraction, leading to a flaccid type of paralysis. Such a flaccid paralysis has been observed to follow the spastic paralysis in nematodes that have been incubated with levamisole for some time. How this relates to recovery of movement by the worms is not clear. Worms that have recovered movement in the presence of levamisole are more resistant than untreated worms to paralysis by pyrantel, bephenium or methyridine. In *A. suum*, the pyrantel-induced contraction is inhibited by piperazine (a GABA agonist) or d-tubocurarine (a nicotinic acetylcholine antagonist), whereas the levamisole-induced one is not (Coles, East and Jenkins 1975; Aubry *et al.* 1970); this suggests that the binding sites for the two compounds are not identical. The binding of the four anthelmintics may be restricted to the interneuronal acetylcholine receptors and may not affect the neuromuscular junctions, because 'recovered' parasites remain sensitive to paralysis by acetylcholine, which probably acts in these 'recovered' worms directly at the neuromuscular junctions.

Acetylcholinesterase is the enzyme responsible for removing acetylcholine at synapses. Many compounds inhibit this enzyme in many organisms; acetylcholinesterase inhibition is the basis of the activity of the organophosphate insecticides, and some of these (e.g. dichlorvos) are also selectively effective against helminths. Most of these compounds, however, are also toxic to vertebrates. The schistosomicide metrifonate, which is rearranged in the host's blood to the active compound dichlorvos, is effective *in vivo* against *S. haematobium* but not *S. mansoni*, although the *in vitro* susceptibilities of their acetylcholinesterases are similar (see Bennett and Depenbusch 1984). However, the acetylcholinesterase of the host is only slightly less susceptible to inhibition than that of the parasite and, since dichlorvos inhibits many other enzymes, selectivity may in fact be due to another effect altogether. Treatment causes temporary paralysis of the worms that lasts for the duration of exposure to the drug. The worms cannot maintain attachment using their ventral suckers and are shifted from their normal sites by the flow of blood. The different susceptibilities of different schistosome species may be due to their different locations in their hosts. *S. mansoni*, if temporarily paralysed in the mesenteric veins, would be shifted to the liver, from which it could return to its normal site after the paralysis had subsided. *S. haematobium*, on the other hand, would be shifted to the lungs on paralysis, from where it is difficult to return to the vesicular plexus of the bladder.

The effect of acetylcholine on nematode neuromuscular preparations is blocked by the anthelmintic piperazine, which causes hyperpolarisation of the muscle membrane, thus preventing depolarisation and causing flaccid paralysis. Piperazine acts directly as an agonist of GABA, which is the inhibitory transmitter in nematode nerves, and it achieves its selectivity because most GABA-ergic nerves in the host are protected by the blood-brain barrier in the central nervous system.

GABA is synthesised in neurons from glutamate, a reaction catalysed by glutamate decarboxylase. It is secreted into the synapse, where it binds to GABA receptors on the post-synaptic membrane. GABA is removed from the synapse by the presynaptic neuron and metabolised to succinic semialdehyde, then succinate, via 4-aminobutyrate aminotransferase and succinate-semialdehyde dehydrogenase. These enzymes and the uptake process are all potential targets for chemotherapy.

The GABA receptor on the post-synaptic membrane is part of an ion channel complex controlling the Cl^- channels. When GABA binds to the GABA receptors, the Cl^- channel opens, causing hyperpolarisation of the post-synaptic membrane, thus preventing signal propagation. The mammalian GABA receptor-Cl^- ionophore complex has many interesting properties, in that it possesses binding sites not only for GABA but also for indirect GABA agonists – the benzodiazepines and avermectins – and for picrotoxin, a non-competitive GABA antagonist (Figure 5.8; see Krogsgaard-Larsen 1981). Barbiturates also bind at the picrotoxin site. There is interaction between binding sites – for example, binding of avermectins increases the binding of benzodiazepines and GABA. Not all mammalian GABA synapses necessarily have the complete set of binding sites.

The potent anthelmintic and insecticide ivermectin (avermectin $B_1 a$) acts against GABA-ergic nerves in nematodes and arthropods (see Campbell 1985). In *A. suum* it causes immobilisation by potentiating the action of GABA at two sites, at the inhibitory neuromuscular synapses and at the junctions between ventral interneurons and dorsal excitatory motorneurons (Kass, Stretton and Wang 1984). By analogy with the effects of avermectins on mammalian brain preparations, ivermectin probably binds to the specific, high-affinity avermectin binding sites on the GABA receptor-ionophore complexes of the post-synaptic membrane, thus enhancing the binding of GABA to the receptor. At the same time, avermectins also increase pre-synaptic release of GABA (see Pong and Wang 1982). Both effects cause hyperpolarisation and prevent signal transmission.

Compounds acting at the GABA receptor complex have a great future in chemotherapy because of their potency and selectivity. The benzodiazepine receptors and also the picrotoxin/barbiturate receptor of this complex are also potential sites for antinematode chemotherapy. It is of interest that the neuromuscular GABA binding site in *A. suum* is not blocked by picrotoxin, which indicates that it is different from the interneuron site.

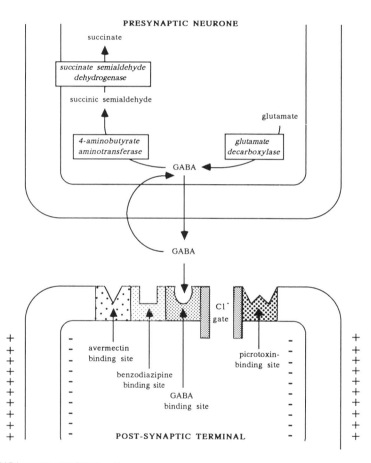

GABA: gamma-aminobutyric acid

Figure 5.8: The Mammalian GABA Synapse and a Diagrammatic Representation of the Receptor Complex

5.4 Conclusions: Adaptation to Parasitism?

In this survey of drug targets in parasites, we have identified many areas of parasite physiology and metabolism that have the potential to provide useful sites for chemotherapeutic attack. Although we have gained some insights into certain specialised features of metabolism in parasites, it is not obvious which of these we can attribute to the parasitic habit, and which are due to the evolutionary origins of the groups or to features that the parasitic habitats have in common with other, non-parasitic, habitats in the biosphere. It is difficult to resolve this question in the absence of parallel studies on free-living members of the parasitic groups (where they are available), but a number of aspects have emerged that we can confidently assign to the parasitic habit. These fall into two groups, (i) loss of metabolic functions that are apparently unnecessary to the parasites since they are provided by the host, and (ii) development of functions specialised for the parasitic state.

Probably the clearest metabolic example, common to all the parasitic groups, is the absence of the purine *de novo* synthetic pathways and the elaboration of the salvage pathway networks. These networks differ between parasite species and the pathways that operate are, no doubt, closely linked to the salvageable entities available from their respective hosts. It is not clear whether the absence of the *de novo* pathway in parasites is a result of genetic loss or of gene repression; this question can be resolved with further study. In either case, the absence of a functioning *de novo* pathway in the parasitic Protozoa and helminths is common to all the phyletic groups and can be considered a clear adaptation to the parasitic habit. By eliminating the *de novo* pathways of purine synthesis, parasites economise on the synthesis of enzymes and control processes that are not essential and rely on host-supplied salvageable substrates. Another example in this category may be the apparent loss of the cytosolic enzymes of the porphyrin synthetic pathway in kinetoplastids, which thus have an absolute requirement for haem from their hosts.

If the 'loss' of the purine *de novo* pathway is an adaptation to the parasitic habit, then it is not at all clear why most parasitic groups have retained the ability to synthesise pyrimidines *de novo*. Many parasites posess both *de novo* and salvage pathways for pyrimidines, so the solution to the problem may lie in the nature of the supply of salvageable pyrimidine bases from the host, or, possibly, in the molecular mechanisms controlling repression of the genes for the pyrimidine salvage enzymes in these parasites.

Other aspects of biosynthetic metabolism appear to be lacking in parasites. Parasitic helminths are unable to synthesise long-chain fatty acids or steroids *de novo* and are unable to desaturate fatty acids. However, many invertebrate groups, both free-living and parasitic, lack these pathways and their absence is not considered to be a specific adaptation to parasitism (see Barrett 1981).

Other metabolic adaptations to parasitism are less easy to identify and are certainly less universal than the loss of *de novo* purine synthesis. In past years, much emphasis has been given in the literature to apparent 'adaptations to parasitism' in the energy metabolism of helminths – the anaerobic pathways, partial reverse TCA cycle activity, fumarate reductase – but it is clear that the operation of these pathways is not necessarily an adaptation to the state of parasitism, but rather to the physicochemical conditions of the environments in which the parasites are found. Similar pathways are found in free-living organisms subject to the same types of environment and were probably also present in the free-living ancestors of the parasitic groups. The specialised anaerobic pathways of *Giardia*, *Entamoeba* and the trichomonads are also probably phyletic in origin and were present in their free-living ancestors. We consider, however, that the retention of active anaerobic pathways by parasites occupying relatively aerobic environments probably is, indeed, an adaptation to the parasitic habit, for reasons discussed below.

An important feature of energy metabolism in many parasites is their reliance on carbohydrate (and, in some groups, glycerol) as an oxidisable substrate, to the exclusion of lipids or amino acids. In many groups, this would be a consequence of low TCA cycle activity and low oxygen availability forcing a greater emphasis on glycolytic reactions, for which glucose is the major substrate. But in other groups, especially the blood-dwelling parasites (schistosomes, filariae, trypanosomes), glucose remains the major substrate and glycolysis the major catabolic pathway despite relatively high oxygen concentrations in their environments. In addition, some groups (e.g. bloodstream trypanosomes) do not even store carbohydrate and are thus dependent on the host's supply at all times. In all these groups the genetic capacity for fully oxidative metabolism has not been lost, because it is functional at other stages of the life cycle. This means that the expression of fully oxidative metabolism is suppressed during the blood-dwelling stages, presumably for economic reasons related to the uninterrupted supply of unlimited glucose in the host's blood.

Many parasitic helminths are unable to oxidise lipids because the parasitic stages do not possess a functional β-oxidation sequence, although most of the enzymes of the pathway are present. Further work is needed to demonstrate whether this is a universal feature of parasitic helminths. A number of free-living stages of nematodes do possess functional β-oxidation sequences, and they rely on lipids as substrates for energy metabolism. Therefore the genetic capacity for β-oxidation is present in these groups but is not expressed in the adult stages.

It could be argued that, in the absence of any significant oxygen-linked electron transport activity, many helminths do not have the capacity to oxidise the substantial amounts of NADH generated by β-oxidation. The absence of oxygen-linked reoxidation of NADH, and, consequently, of β-oxidation, would therefore be an adaptation to the specific environmental conditions in which these parasites live. But NADH can be oxidised anaerobically via the fumarate reductase reaction, so low oxygen levels are not necessarily a barrier to catabolism of lipids via β-oxidation in parasitic helminths. On the other hand, as discussed above, parasites may be supplied with more than adequate supplies of carbohydrate substrates and therefore have no need to synthesize the enzymic and control machinery required to employ β-oxidation as a source of energy. In this case, loss of the β-odixation sequence could be considered a specific adaptation to parasitism.

It is interesting to consider the case of the mitochondrial glycerol-3-phosphate oxidase system in bloodstream *T. brucei* and the presence of glycosomes in all the kinetoplastids. These parasites essentially use molecular oxygen to reoxidise glycolytic NADH, without deriving any additional ATP from the process. High ATP levels are maintained by an extraordinarily rapid rate of glycolysis. This highly wasteful utilisation of glucose is unlikely to occur in free-living organisms and could only be tolerated by an organism with unlimited supplies of glucose and oxygen. For the same reason, these parasites do not maintain stores of oxidisable substrate such as glycogen. In the absence of any information about free-living members of this group, it seems reasonable to conclude that the operation of these pathways in bloodstream *T. brucei* is an adaptation to the parasitic habit. Whether this argument can be extended to the presence of glycosomes in these organisms is an open question. These organelles concentrate the enzymes of glycolysis and may have the function of chanelling substrate flow in order to permit extremely high rates of flux through the pathway. This would clearly be advantageous in organisms requiring high rates of glycolytic flux. There is no evidence at present from free-living relatives or evolutionary ancestors to show whether glycosomes predated the parasitic habit or provided a useful 'preadaptation' for parasitism. The presence of glycosomes only in the kinetoplastids points to a possible 'preadaptation', or evolutionary accident that may have proved advantageous for an evolving parasite.

It is quite clear that a major and essential area requiring quite precise adaptation in parasites is the external surface of the organism that is in close contact with the host. The parasite surface contains molecules and receptors that interact directly with host-derived molecules in a highly integrated fashion. Surface molecules function as digestive enzymes, transport channels, receptors, and as recognition and adhesion sites, for example. These are all precisely adapted for the function in question, and many would be specific for a particular host-parasite system.

Surface molecules on internal parasites are also antigens accessible to the immune system of the host. Particularly interesting examples of adaptation are seen in the immune evasion mechanisms employed by some parasites – for example, the production of large quantities of antigenic molecules that act as 'decoys' or 'smokescreens', the production of a protective and disposable outer coat that has not significant metabolic function, the production of variable populations of antigens, and mechanisms for adsorbing or even synthesising host molecules that prevent the immune system from 'seeing' the invader. Other internal parasites appear able to cause immunosuppression in their hosts.

The massive effort and resources put into reproduction by all parasites is closely linked to increasing the probability of finding the next host to perpetuate the life cycle. This is not necessarily an adaptation specifically to parasitism, but rather a consequence of inhabiting a 'patchy' environment, in which suitable habitats (hosts) are not common or evenly distributed. Free-living organisms inhabiting 'patchy' environments employ similar reproductive strategies. However, parasites are in a position to devote a greater proportion of their metabolic resources to reproduction than free-living organisms because their hosts provide not only many of the resources but also spare the parasite's consumption of these resources by providing energy-consuming functions such as protection and dispersal mechanisms.

From this discussion it is clear that chemotherapy takes advantage of a variety of differences between hosts and parasites, but that the 'differences' are not necessarily adaptations to parasitism.

6

Biochemical Variation in Parasites

6.1 Introduction

In an analysis of mutational change in fossil organisms, Schopf (1982) comments that paleontologists recognise that the skeletal changes that can be observed in 'hard parts' in any particular evolutionary lineage probably underestimate the total change in the organism by a factor of ten. This is almost certainly far too conservative. As well as the hard parts, there are the 'soft parts' to consider and, most important, there are functions. Metabolism, physiology, development and behaviour all have genetic bases and all are subject to mutational change.

The genome of an organism can conveniently be considered in two major parts. The first is the part that is expressed, generating the visible attributes of the organism, the phenotype. It comprises a series of DNA sequences that, when activated, act as templates upon which are assembled complementary strands of mRNA (messenger RNA). The mRNA then directs the assembly of proteins, many of which are enzymes, whose complex interactions produce the organism. The second part of the genome is not expressed. It contains the so-called redundant DNA plus DNA with a regulatory function which controls the expression of the rest of the genome. The genome itself interacts with a complex intracellular environment containing regulatory proteins and other macromolecules, low molecular weight cofactors and inorganic ions.

Mutations may occur either in that part of the DNA that is expressed, with consequences for the structure and function of the organism, or in the regulatory DNA, which may affect development, for example. They may also occur in the 'redundant' DNA and have no immediate or overt effects at all, except to act as a repository for unexpressed mutations that may, under certain conditions, provide an internal source of variation for the organism. The genome may also interact with itself. If, for example, a gene coding for a regulatory histone is altered, it may alter the *milieu* of the genome, changing its mode of regulation, with far-reaching results. There are few examples of post-transcriptional regulation   of gene expression in parasites but an exciting one has recently been reported in *Hymenolepis diminuta* by Siddiqui, Karcz and Podesta (1987). Their study of RNA metabolism suggests that the regulation of gene expression during adult development and sexual differentiation in this tapeworm occurred almost exclusively at the post-transcriptional level. Unfortunately, the the actual mechanisms by which this occurs remain obscure.

It is important to be able to recognise true genetic variation, arising from differences in the genome, and to distinguish it from phenotypic variation that arises from interaction with the environment. Physical or chemical factors in the environment of an organism may interact directly with cellular or sub-cellular factors. If a component in the environment varies, such as a persistent or cyclic change in temperature, redox conditions or the presence or absence of an important molecule, then the intracellular environment may be altered, affecting the interactions of the components of the cellular sytem. There are a number of well established examples of this type of variation induced by the external environment. Perhaps the best known is that of *Daphnia* species that inhabit fresh water lakes in North America (Pennak 1978). They undergo a process known as cyclomorphism. This cyclic phenomenon is a response to the seasonal temperature changes in the environment. In the cooler seasons, spring and autumn, they are normally round headed. In the summer, however, as the water temperature rises, their heads become progressively more elongate. This is not a stable polymorphism – the condition where two or more distinct genetic types are maintained within a population of a single species – although the unwary investigator might erroneously conclude that there are two or more different morphs. Similar examples can be found among some trematodes, whose exact pathways of development depend on the prevailing temperature. Thus, in Kenya, *Fasciola gigantica* includes rediae, but not metacercariae, in its life cycle in cool conditions. When the temperature rises above 16°C, however, both rediae and metacercariae are produced alternately (Dinnick and Dinnick 1964).

These examples could be considered to be extreme manifestations of variation resulting from slight changes in normal developmental 'cues' that are present in the organism's normal environment. Similar cues could be found in the life cycle of almost any parasite or free-living organism that is investigated, though in the majority of cases the precise nature of the cue has not been identified. Good examples of such effects are the changes in morphology and metabolism of the *Trypanosoma brucei* group during its blood stage in vertebrates, where long-slender forms change to intermediate and short-stumpy forms, and in the transformations that occur when these forms are either ingested by the insect vector or experimentally placed in culture. Although this is 'variation', it is part of a normal developmental process and must be recognised as such and distinguished from true genetic variation. Experimentally, this may not be easy, particularly in the case of the Protozoa, and it is necessary in many cases to use comparative DNA analyses to confirm the existence of stable genetic variants within a species.

Independent and stable variants must also be clearly distinguished from stable genetic polymorphism within a population. The latter occurs quite commonly in nature, when an isolated population consistently produces two or more distinct varieties. In a stable population breeding randomly, the frequencies of these varieties are found in a Hardy-Weinberg equilibrium and the varieties are maintained in that population by consistent, often cyclical, selection pressures. A good example is sickle cell anaemia. In certain human populations in East Africa the gene for an abnormal haemoglobin, haemoglobin S, is quite common in the population. Haemoglobin S is structurally unstable, tends to polymerise when deoxygenated and causes collapse or 'sickling' of the red blood cells in which it occurs. Individuals homozygous for this gene, having received it from both parents, have abnormal red blood cells, suffer from sickle cell anaemia and usually die in childhood. Individuals heterozygous for this gene, having received it from only one parent, have a small proportion of sickle cells in their blood, and sufficient normal ones to permit their survival. Such people have increased resistance to infection by malaria, in comparison with individuals homozygous for one of the many normal haemoglobins. This means that heterozygotes have a selective advantage over those homozygous for normal haemoglobin, because a proportion of the latter die in childhood from malaria. Thus, the polymorphism is maintained so long as lethal childhood malaria is present to act as a selecting agent. Hence, the variants in cases of genetic polymorphism are not independent of each other and their proportions in the population remain stable only while under a specific selection pressure. Genuine cases of genetic polymorphism are not always as easy to recognise as the case of sickle cell anaemia; in many cases the relative gene frequencies deviate from the Hardy-Weinberg equilibrium and the selection pressures are unclear.

If a polymorphism in a population is perturbed in some way, usually by a change in selection pressure brought about by extraneous factors or by isolation, it is likely that one of the 'morphs' will disappear, or that different morphs from the original population will survive in different locations. In these cases a genetically stable variant population will emerge as gene flow from the general gene pool has been effectively 'bottle-necked' and a new strain will be established and will be maintained – provided there continues to be no gene flow between this population and adjacent ones. Such a scenario is quite probable for parasites when they become established in new hosts or are cultivated in laboratories. Each time this occurs, it provides an example of isolation in which gene flow has been restricted and, at the same time, selection pressure has changed.

6.2 The 'founder' principle

In a bottle-necked population, the capacity to produce all the morphs present in the original population is severely restricted, because the complete range of genotypes from this population is no longer represented. The bottle-necking of a population can be brought about by a number of mechanisms, including isolation of a small fraction of the original population. For parasites, this could include movement of part of the host population to a new environment, arrival of a new host population or species, death or removal of a large fraction of the host population – or a significant change in selection pressure such that only a small proportion of the original population survives.

When a parasite finds itself in a new host – this could be a new host species, or a subpopulation of the same species with different physiological or behavioural attributes – the successful colonisers represent only a small fraction of all available genotypes for that particular population. This sample is further reduced as the new environment – that is, the new host – selects out genotypes that are not viable. Those remaining are therefore quite unrepresentative of the population from which they were derived. The parasites themselves often intensify this effect by a process of genotype amplification, that is, asexual reproduction or cloning. Protozoa such as *Plasmodium* and *Toxoplasma* species are well-known to establish clones readily, but it is perhaps not so evident that some platyhelminths can also establish clones by incorporating asexual reproductive stages in their life cycles. The larval stages of *Echinococcus granulosus, E. multilocularis, Mesocestoides corti* and many trematodes provide good examples of such amplification. Provided the new clones are viable at all stages of the life cycle, a successful subpopulation may arise that is genetically distinct in certain respects from the original population. It remains distinct so long as there is no genetic mixing with the original population and (or) the new host population continues to be available.

When parasites are isolated from the field and established as cultures in laboratories, the problem of gene-pool bottle-necking is particularly severe, since generally only a small fraction of the original field population is sampled. In addition, the parasites must adapt to the specific conditions prevailing in a given laboratory. Many protozoan parasites are cloned, thus reducing their gene pool dramatically. Many parasites are cultured in host species different from those available in the field, or in inbred host species; they are not subject to the same environmental conditions that occur during any natural free-living phase of their life cycles; intermediate or vector stages may be by-passed; they may be frozen in liquid nitrogen for long periods of time. The combination of gene-pool restriction and intense selection pressure results, not surprisingly, in the appearance of strain differences in species of Protozoa and helminths held in different laboratories. The ultimate cultivars may be as representative of the wild type as the chiahuahua is of the genus *Canis*.

The extreme case of laboratory cultivation illustrates very clearly the 'founder principle' at work in the establishment of new strains of parasites. This is a concept introduced by island biogeographers to explain the often unique assemblages of flora and fauna encountered on islands that are isolated by considerable distances from the mainland. Although elaborate mathematical models can be constructed to predict the numbers of species to be encountered and to show relationships between these numbers and distance from sources of colonisation, they cannot predict the nature of the successful colonising species because that is a result of purely stochastic processes. On islands, therefore, the gene pools reflect the genetic makeup of those organisms that had the good fortune to arrive first and survive.

There are some obvious analogies between islands and the hosts of parasites. Hosts are isolated in space and many of the most elegant adaptations of parasites are devoted to solving the problem of getting the next generation to the next host. Except in cases of intense and repeated infection which probably only occurs frequently in extensive monocultures, such as flocks of sheep, parasites within their hosts are usually isolated from the gene pool of the population until the next generation. A parasite that finds itself in an unusual host is likely to be permanently excluded from contact with the gene pool. Parasites able to survive and reproduce in such a situation may found a new strain 'adapted' to a new host.

In addition to restriction of the parasite's gene pool resulting from forms of isolation, parasites may be subject to intense changes in selection pressure in the field. A good example can be found in the use of antiparasitic drugs and the development of strains of parasite resistant to such compounds. If treatment with a given concentration of a drug does not remove 100% of the parasites that were exposed to the drug, then the survivors probably have some resistance to the compound in question. The genes conferring resistance, normally present at a very low level in the parasite population, thus give their possessors a selective advantage when a given antiparasitic compound is used. Since the survivors con-tribute those genes to the next generation, the frequency of genes conferring resistance increases with each cycle of selection, and a resistant strain may develop. The rapidity with which such a strain develops, and its stability, depends on the intensity of selection, the overall and relative fitness of the resistant individuals, and the extent of dilution of the genes in the general (unexposed) population. The question of resistance to parasitic compounds is discussed in greater detail later in this Chapter.

6.2.1 Metabolic variants

Metabolic pathways are the means by which the gene products do their work in organisms. They can vary in a number of ways. The first, and perhaps the simplest to understand, is the variation that occurs in response to changes in the local environment of individual cells. This metabolic regulation includes feed-forward and feed-back control of enzyme activities, enzyme activation or inhibition by substances intrinsic or extrinsic to the pathway concerned, increased or decreased rates of synthesis or degradation of enzymes. In this way, the output from a metabolic pathway can vary according to the requirements of the organism at given time.

Much traditional biochemical research has been concentrated on this important aspect of metabolism in metazoan organisms. Thus, a liver contains pathways, among many others, for the synthesis or degradation of glucose or glycogen, for respiration and for maintaining redox balance. The activities of all these pathways are finely tuned between limits that can be exceeded only to the detriment of the organism. A biochemist who studies mammals confidently expects that all normal rat livers – and possibly also livers from other rodent species – will conform to the same general pattern of metabolic pathway, in terms of constituents and control. The possibility has not been seriously entertained that Wistar rats and Hooded rats, for example, might be sufficiently different to possess generically different metabolic pathways. This is somewhat surprising because other workers on mammals, the immunologists, have long recognised strain variation in their favourite organisms and employed it as a valuable research tool (e.g. Mitchell 1979a, b & c). Indeed, there is a small amount of evidence for biochemical variation between mouse strains, for example. Investigations of the enzymes of pathways of purine metabolism in blood cells of different strains of mice have shown quite significant differences between strains (Henderson, Zombor, Johnson and Smith 1983). It would not be surprising to find similar differences in other metabolic pathways.

Biochemists who work on microorganisms have quite a different view. In a given species of bacterium or yeast, some entire pathways may not be active at a given time. They are induced only in response to specific environmental circumstances. In fact, some pathways may not be represented at all within a reproductive cycle if conditions are not appropriate. Variation thus takes on another dimension. These organisms possess the regulatory mechanisms described above but, in addition, they are able to elaborate different metabolic pathways in an opportunistic way. In bacteria and other microorganisms, then, metabolic pathways may be variable, in that they may be present or absent. It is also possible that they are active only in part. All these effects may be controlled by environmental conditions.

6.3　Parsimony, conservatism and optimisation

For a long time there has been a feeling amongst biological scientists that nature ought to be parsimonious. This follows from the 'adaptationist' view of organisms. A natural law of parsimony has never been formally stated but could probably be phrased as follows. An organism will carry out a given function in the most economical way consistent with the constraints imposed upon it by both the properties of its component parts and the total functioning of the organism as a unit. This implies that once a thermodynamically efficient way of carrying out a process has evolved, that process will be conserved. Unfortunately, there is no unequivocal evidence for such parsimony in nature, and recent work suggests that this is certainly not the case in evolution (Johnson, 1982). So, while some aspects of biology may appear parsimonious, others appear spectacularly extravagant. Or – is it that they seem that way because we do not have all the facts? Parsimony, as a general principle, cannot be recruited to help us to understand organisms. Some phenomena are parsimonious and lead to conservatism – the active centres on enzymes may be conserved through epochs – but others are not.

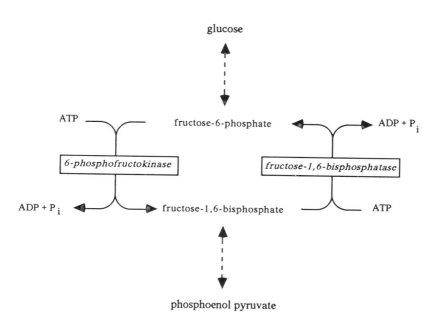

Figure 6.1: An Example of a 'Futile' Cycle

A good example of such apparent extravagance in biochemical evolution is the occurrence of 'futile' cycles in some metabolic pathways. These cycles occur at points where two or more enzymes catalyse essentially irreversible reactions in which the product of one is the substrate for the other(s). Usually one enzyme of the cycle utilises ATP, and the reactions may involve phosphorylation and dephosphorylation, methylation or degradation. Examples of such enzyme pairs include hexokinase (glucokinase)/glucose-6-phosphatase, glycogen phospho-rylase/glycogen synthase and 6-phosphofructokinase/fructose-1,6-bisphosphatase (Figure 6.1). If each enzyme of the pair operates at the same rate, ATP is consumed but there is no net flow of carbon. The discovery of these futile cycles posed conceptual problems for biochemists because they appeared to demonstrate imperfect control over enzyme activities. It was considered that the activities of these enzymes would have to be very tightly regulated in order to prevent such futile cycling and apparent dissipation of ATP. Experimental evidence has shown that these futile cycles are indeed active in organisms, though normally at low rates, and that the cycling rate varies with physiological conditions. Is this 'imperfect control' – anathema to adaptationists – or do these cycles have an important metabolic function? Two major functions have now been attributed to futile cycles: they may be responsible for the generation of heat and they may act as very sensitive control points in pathways requiring very fine tuning and rapid changes in activity (Koshland 1984; Newsholme, Challis and Crabtree 1984). Thus, extravagant consumption of ATP, as demonstrated by 'futile' cycling may have functions not immediately obvious to the investigator: this problem considerably confuses the debate about parsimony.

All parasites probably possess some of the enzyme pairs responsible for futile cycles, and cycling at 6-phosphofructokinase has been demonstrated in living *F. hepatica* (Matthews, Foxall, Shen and Mansour 1986). The parasitic stages of most parasites would probably not need active futile cycles if their primary function were thermogenic, since an appropriate body temperature would normally be supplied by the host. Therefore, futile cycles operating in such internal stages of parasites would be expected either to have other functions, such as sensitive metabolic control, or to be a sign of imperfect control or 'redundancy'.

Another example of apparent lack of parsimony in biochemical evolution is found in those parasites, both helminth and protozoan, that live in the blood of vertebrates. These organisms live in an environment containing sufficient supplies of oxygen to support aerobic energy metabolism, yet many of them exhibit either complete (*Plasmodium berghei*) or partial (*Schistosoma mansoni, Litomosoides carinii, Dirofilaria immitis*) anaerobic metabolism, or modified aerobic metabolism with highly diminished yields of ATP (*Trypanosoma brucei*). Why should this be so, when aerobic catabolism of glucose, probably the principle respiratory substrate in these parasites, yields as much as 10 times more ATP than anaerobic catabolism? It could be argued that the ability to oxidise glucose completely aerobically would confer a selective advantage on such organisms since they would utilise their substrates more economically. But, since this is not the case, either they are not fully 'adapted', or there are considerations other than energy yield involved. Glucose is unlikely to be in short supply in the blood of most vertebrates, so perhaps 'economical' utilisation of an unlimited substrate is not a selection factor. In addition, there may be costs associated with fully aerobic metabolism, which requires specialised mitochondria, electron transport systems, oxygen transport, control mechanisms, that we are unable to estimate. Possibly some parasites do not possess the genes for the components of aerobic metabolism; but this is not the case for the three helminth species cited above – in each case at least one stage of the life cycle has aerobic or partially aerobic energy metabolism. There may be barriers for the diffusion of oxygen, or oxygen levels may fluctuate such that partial aerobic metabolism, which retains some flexibility, is the most appropriate. Whatever the explanation, it is clear that these parasites have in most cases the genetic capacity for oxidative metabolism, but that it is not fully utilised. To the extent of our present knowledge, this does not appear to fit the concept of parsimony in evolution.

Since it appears, therefore, that organisms are not necessarily parsimonious, there should be no conceptual barrier to the proposal that metazoans, as well as bacteria, are able to exhibit considerable metabolic variation within species, and that metabolic pathways in these animals may vary without violating some immutable law. There are, however, limitations to what is possible. If an organism is considered as a regulatory control system, presumably it is regulated so that the sum of its activities is *optimised*. An optimised system is one of compromise; therefore, an apparently less efficient pathway may be utilised because other functions or factors impose demands which that pathway alone can satisfy. The demands of reproduction in parasitic helminths, for example, may require substantial lipid storage in the adults, despite the thermodynamic imperative that lipids be used immediately as the most energetically efficient source of ATP. Another example is the bloodstream form of *T. brucei*, which enjoys a constant supply of glucose and possesses glycosomes which permit very rapid rates of glycolysis. This allows such rapid rates of ATP generation that the costs associated with development of fully oxidative metabolism are unnecessary at that stage of development. A *maximised* system, on the other hand, in which the most thermodynamically efficient process consistent with the resources available is always favoured, lacks *flexibility*. Successful survivorship demands the husbanding of resources and their optimal allocation to different functions. Retention of flexibility is important to any organism or population.

6.4 Experimental detection of variation

Variation in parasites is detected by a number of techniques, biological or biochemical, some of which have become standard tools for the geneticist or taxonomist. For many years it has been known that parasites belonging to (apparently) the same species could vary in virulence, in rates of development, in antigenicity, in host specificity, in susceptibility to chemotherapy, and in many other ways. Some of the genetic bases for this variation are now beginning to be revealed (see, for example, Walliker 1983a), and basic questions can be posed about the significance of such variation in parasites. In this section we will consider some of the major biochemical techniques in current use and how the results contribute to our understanding of variation in parasites.

6.4.1 *Electrophoresis and isoelectric focusing*

Enzymes and other proteins may be solubilised and subjected to electrophoresis, in which they are separated by net charge, or to isoelectric focusing, in which they are separated by charge in a pH gradient such that they move no further once they reach their isoelectric point. Separated proteins are detected and identified by specific staining methods. This technique thus detects variations in proteins where a change in net charge has occurred, but it is limited to proteins that can be readily stained and to those that retain their activity under the experimental conditions. It has proved very useful in taxonomy because large numbers of enzymes can be surveyed – thus representing large numbers of genes of known function – and cases of polymorphism can be used to characterise populations and subpopulations (zymodemes). Isoenzymic characterisation of a population permits estimation of genetic variability within the population, calculation of gene flow within and between populations of the same species, calculation of genetic identity and genetic distance, and of similar parameters that allow taxonomists and geneticists to allocate organisms or groups of organisms into their appropriate species and strains. Knowledge of the genetic relationships within and between populations of parasites has made a major contribution to our understanding of the epidemiology of the diseases they cause, especially in the case of some parasitic Protozoa such as *Trypanosoma cruzi* and the *Leishmania* groups.

Lactate dehydrogenase is a classic example of an enzyme exhibiting polymorphism in almost every organism that has been examined. In mammals, for example, this enzyme is a tetramer of four units, each with a molecular weight of 35,000. There are two types of units – A and B polypeptides. Five combinations of subunits are possible and are, indeed, found: AAAA, AAAB, AABB, ABBB and BBBB. In this case, the different isoenzymes have different kinetic properties, their distribution varies between different tissues and the kinetic properties can be related to the functional requirements of the relevant tissue.

For example, B-containing enzymes predominate in aerobic tissues like the mammalian heart. One explanation is that B has a greater affinity for NAD^+, permitting lactate oxidation. Lactate does not therefore accumulate in heart muscle, a strictly aerobic tissue. On the other hand, skeletal muscle is capable of sustaining bursts of anaerobic activity and is tolerant of lactate. The A-containing enzymes bind NADH, permitting the reduction of pyruvate, and lactate accumulates (Dixon and Webb 1979).

Isozyme patterns can be much more complex than this, however. In North American lake trout up to fifteen isozymes of lactate dehydrogenase have been detected, apparently due to hybridisation between adjacent species or subspecies. Each taxonomic unit has its own characteristic electrophoretic pattern of five isozymes – it forms a zymodeme. During hybridisation the offspring receive genes from two taxonomic units, giving rise to a large number of varieties. It is not clear whether these varieties are in any way functionally different from each other, so the adaptive significance of these observations cannot at present be assessed (Goldberg 1966; Wuntch and Goldberg 1969).

Enzyme polymorphism in parasites
The application of electrophoretic methods to parasites is complicated, for many reasons. First, the parasite must be recovered intact, free if possible from

contaminating host proteins. This presents problems in the case of very small parasites, or those dwelling intracellularly or within tissues, or parasites with guts containing host material. Second, it is not generally possible with parasites, except the largest of the helminths, to dissect out single tissues. Since different isozymes of certain enzymes may be found in different tissues, results obtained from homogenates of whole organisms may be difficult to interpret. Third, the developmental stage of the parasite needs to be considered. Different developmental stages often have different isozymes. Adult helminths contain eggs or larvae which may have their own pattern of isozymes, and which may be the products of cross-fertilisation with other individuals. Adult cestodes contain a range of developmental stages, plus eggs, all possibly represented by different isozymes. Fourth, effects of the host or external environment must be recognised so that normal metabolic variations – *phenotypic* variations – are not construed as genetic or genotypic variation. This may be difficult to assess because different hosts, for example, may create an environment in which different gene loci are expressed from the same genotype. For the same reason, it is important to use recently-isolated field strains for epidemiological studies rather than laboratory-cultivated ones. Fifth, different strains of the same parasite species may coexist within a single host. If the parasites are small, so that groups rather than individuals have to be assessed, interpretation may become difficult unless each strain has been characterised separately and good markers are available. Despite these limitations, some good results have been achieved in applying this technique to parasites.

The complex taxonomy of *Leishmania* species has confused workers for many years. This genus was considered to be undergoing speciation in Central and South America and the relationships between the New World genera and Old World ones were unclear. The initial classifications and identification methods used criteria such as epidemiological differences, developmental studies in sandflies, growth rates *in vitro* and in non-human hosts, morphology, serotype and DNA buoyancy. These criteria suffered from lack of reproducibility (due partly to variations between strains of experimental hosts) and overall complexity. Recent work using protein polymorphisms (see Miles 1983, for review) has elucidated relationships among and between Old and New World forms and shown, for example, that *L. major* and *L. tropica,* previously thought to be closely related subspecies, are distinct from each other and restricted to the Old World, whereas *L. donovani infantum*, a Mediterranean species, is responsible for visceral leishmaniasis in the New World, presumably imported with the Spanish invaders. In addition, it has been demonstrated that *L. mexicana amazonensis* is homogeneous with respect to protein polymorphism over a wide geographical area of Brazil, yet it is the agent for both the simple cutaneous human infection and the incurable diffuse type of infection in that region; the difference is ascribed to the immunocompetence of the infected host. Such studies are extremely important in the design of control and treatment programmes for human leishmaniasis.

Trypanosoma cruzi populations in South America consist of many highly variable 'strains', originally characterised by geographical locality, infectivity, virulence and pathogenicity (Brener 1973). These biological parameters were not sufficiently reliable to distinguish the different strains consistently. Three separate groups (zymodemes) of *T. cruzi* have been identified by enzyme analysis; they represent three distinct types of transmission cycle and are designated Z1 (sylvatic cycles), Z2 (domestic cycles) and Z3 (infecting armadillo and occasionally humans) (see Miles 1983). The chronic human disease is caused by Z2. Parasites from Z1 or Z3 have been shown to cause sporadic acute cases of Chagas' disease in man in the Amazon basin where Z2 does not occur. The calculated genetic distances between these zymodemes are sufficient normally to warrant taxonomic separation into subpecies or species, but this has not yet been done for *T. cruzi* because the amount of genetic exchange between the units is uncertain and because other biological parameters are not resolved. The organisms appear to be diploid, because heterozygotes are found in some geographical areas. The presence of these heterozygous populations (of uncertain relationship to Z1, Z2 or Z3) correlates with temperate and more variable environments; possibly greater genetic or phenotypic variability is favoured under these conditions.

Studies of enzyme polymorphism have also assisted taxonomic and epidemiological work with helminths (reviewed in Bryant and Flockhart 1986). Two forms of the nematode *Parascaris equorum*, designated *univalens* and *bivalens*, have been shown by this technique to belong clearly to two separate species, *P. equorum* and *P. univalens*; their morphological similarity is presumably due to common ancestry and occupation of similar habitats. *Ascaris lumbricoides* and *Ascaris suum* are genetically more closely related than the *Parascaris* species and are considered to be incipient species. Comparative studies with ascarid species have shown that genetic variability, as measured by mean heterozygosity at enzyme loci, is much less in ascarids with a single host (*Ascaris, Toxocara, Ascaridia*) than in those with multiple hosts during their life cycles (*Anisakis*). This is consistent with a requirement for adaptation to a more variable life cycle which includes a wider range of environments.

It has long been known that many varieties of *Trichinella spiralis* exist. The evidence has not been morphological, but has derived from studies of infectivity and cross-breeding, correlated with geographical location. Isozyme studies by Flockhart, Harrison, Dobinson and James (1982) have shown differences between the temperate domestic strain of *T. spiralis* and tropical and arctic sylvatic isolates. *Onchocerca volvulus* also has several strains. Flockhart, Cibulskis, Karam and Albiez (1986) compared seven enzymes from parasites from four different African countries, and found that four were variable. The variability was such as to enable the researchers to distinguish between forest and savannah isolates.

The human schistosomes are particularly complex – or they may just seem so because of the relatively large number of studies that have been carried out on eleven different species. Detailed studies of 22 populations of Old and New World *Schistosoma mansoni* (reviewed in Bryant and Flockhart 1986) support the view that the New World parasite was introduced with the slave trade, as the differences are slight. On the other hand, *S. mekongi* and *S. japonica* are quite distinct, differing at about 90 percent of the gene loci examined. Calculations of genetic distance suggest that the two species diverged about 10 million years ago. *S. japonica* itself is quite variable; four geographical isolates differed in 17 to 36 per cent of the loci tested.

Echinococcus granulosus, the dog tapeworm that is responsible for hydatid disease in humans and animals, is a particularly famous example of strain variation (Thompson and Lymbery 1988). Several strains exist that can be distinguished on the basis of their host preferences. In England there are two strains, a 'horse' strain and a 'sheep' strain, named for the preferred intermediate host. They seldom – if ever – cross-infect. Enzyme electrophoresis of cyst extracts shows that they differ in eight out of nine enzymes tested.

Are enzyme polymorphisms adaptive?

Enzyme polymorphism is clearly widespread in parasite populations and is useful in taxonomy. The question remains, however, whether such polymorphisms have adaptive significance to the parasites. Some enzymes occur as isozymes with different *in vitro* kinetic properties that would clearly affect the function of the enzyme *in vivo*. Such enzymes may have different distributions between tissues or developmental phases. But is this the case for all polymorphic enzymes? Are they all 'adaptive'? This is very difficult to assess. For an enzyme polymorphism to be considered 'adaptive', a number of general criteria need to be satisfied (Clarke 1975; Koehne 1978).

First, the range of genotypes must give rise to a diverse set of phenotypes whose origins depend upon the different functions of the different isozymes in question. The best that can normally be achieved experimentally may merely be a strong correlation between the two; in some cases no phenotypic diversity (other than the presence of the isozymes) may be detected. This is not necessarily evidence of absence of phenotypic diversity, however, because such diversity may become evident only under specific conditions or developmental stages that we do not or cannot investigate.

Second, the different functions of the isozymes in question must be shown to be relevant to the metabolic economy of the organism. Third, a connection must be demonstrated between a specific gene, its product (the enzyme in question), and a component of the environment. There are few examples of this in the literature. Probably the best are enzymes such as lactate dehydrogenase in lake trout, whose polymorphisms vary with temperature acclimation (Wuntch and Goldberg 1969). Unfortunately, even with these cases, the relationship is one of correlation and not necessarily cause and effect.

Finally, the phenotypic differences must confer some increased fitness. In most cases of enzyme polymorphism this appears to be almost impossible to demonstrate.

The great difficulty in meeting these criteria has fuelled the theoretical evolutionary debate between 'neutralists', such as Kimura (1983), who claim that enzyme polymorphisms are not necessarily adaptive and 'selectionists', such as Ayala (1983) and many others (see Nevo 1983, for review) who claim that they are. The neutralist position is based largely on the twin concepts of genetic drift and the founder effect, by which the genome of an isolated population suffers random variation as a result of isolation. Genetic polymorphism is not considered significant and is simply 'noise' in the system. The selectionist approach, on the other hand, emphasises the role of selective mechanisms in maintaining genetic diversity and enzyme polymorphism. The dichotomy between these two extreme views is artificial and both neutral and adaptational changes probably take place in the molecular evolution of organisms. In general (see Nevo 1983) it has been shown that species occupying more heterogeneous environments are more genetically diverse than those in (by our judgements) more homogeneous environments. This suggests that genetic diversity is not random and therefore likely to be a result of selection. The real question is: how much of each? Considerably more work is required to elucidate this problem and it is important that much of this be done where possible in field models rather than laboratory ones, many of which have become remote from the field situation. In this context it is encouraging to note that Lavie and Nevo (1982) recently sampled two species of marine gastropods and showed that the polymorphism of phosphoglucose isomerase was related to survival in the presence of heavy metal pollution.

An important practical problem in taxonomic assessment of data from enzyme polymorphism studies in parasites is to decide how many such polymorphisms are required to define a separate strain, or to designate a separate species. Could some enzyme polymorphisms be more important than others in this context? If stochastic events alone are responsible for the appearance of isozymes, then polymorphisms will appear from time to time within an otherwise homogeneous population. Selection and stochastic events will then operate and possibly fix the polymorphism in question in a particular population, perhaps in one separated geographically or in a different (but related) host. At that time, other distinguishing characteristics of morphology, development, metabolism, behaviour and host choices may appear. Until these additional changes take place, however, the argument for the existence of separate strains cannot at present be considered convincing. In this context, it is interesting to note the work of Baverstock, Adams and Beveridge (1985) on the *Progamotaenia festiva* complex, a group of cestodes parasitic in macropods. Cestodes from different macropod genera have no distinguishing morphological features yet show very marked enzyme polymorphism. Calculated genetic distances from the electrophoretic evidence would normally imply the presence of a number of well-separated species, but morphologically they are identical. *Practical* decisions obviously need to be made in cases such as this and the concept of separate species or strains in some instances needs to be qualified as much by the techniques used to define them as by biological or evolutionary considerations.

6.4.2 *Nucleic acid analyses*

Genetic analysis of parasites by conventional methods has been difficult because the classical techniques of hybridisation or progeny analysis using mutants or naturally occurring markers are difficult to apply to organisms that have complex life cycles, are difficult to culture and lack easily-recognised genetic markers. An exception is the free-living nematode *Caenorhabditis elegans*, whose genome has been extensively studied and will serve as a useful model for future genetic studies in parasitic nematodes. Some success has been achieved with genetic studies in Protozoa; for example, the genetic bases for chloroquine and pyrimethamine resistance and variation in virulence have been investigated in *Plasmodium* species (see Walliker 1983b, for review) using classical techniques. Other protozoan groups have also been examined, but very little work has been possible with helminths, for the reasons outlined above.

The recent development of recombinant DNA technology has revolutionised the study of genetics and variation because it permits direct study of the genome without the requirement for phenotypic markers. This is a powerful tool that will yield very valuable results in studies of parasite genetics, variation, development and taxonomy. In addition to the standard hybridisation and buoyant density comparisons used to investigate the DNA of different species or strains, it is now possible to compare genomes directly and to probe them for specific genes or base sequences. This can be done using a number of general methods:

(i)   Preparation of a genomic library by digestion of nuclear DNA with a specific restriction endonuclease followed by electrophoresis to separate the fragments. A representative size range of fragments (calculated both to represent as much of the genome as possible and to be of suitable size for insertion) is inserted into appropriate vectors such as lambda phage particles, which may be stored, and which are used when required to infect bacteria and produce clones containing amplified copies of the DNA pieces. Such genomic libraries serve as references for other work and can be probed for specific DNA sequences. It is not usually practical, because of the large size of the nuclear genome, to use such a library for direct comparison between genomes.

(ii)   Digestion of nuclear DNA with restriction endonucleases followed by electrophoresis and ethidium bromide staining of the DNA fragments. This technique detects only the highly repetitive sequences in the genome, such as those coding for ribosomes and histones. Single-copy DNA is not present in sufficient quantities to be detected by staining methods after separation on a gel, because single bands cannot be recognised. Thus usually only a small number of distinct bands from the repeated DNA sequences are seen on the gels; if differences are found, these bands may be used to distinguish species or strains.

(iii) Preparation of mitochondrial DNA, digestion with specific endonucleases followed by electrophoresis to separate the fragments. The mitochondrial genome, which is small (50 kilobases) can be compared directly with mitochondrial genomes from other species, isolates or strains. Since this genome contains a low number of genes, many of which have been identified, it may be easier to identify the actual genes responsible for any differences detected. The mitochondrial DNA is inherited maternally in many organisms.

(iv) Extraction of messenger RNA, preparation and electrophoretic separation of DNA copies to form a cDNA library which is stored in lambda phage and can be cloned in the same way as the genomic library. This library can be used to probe the genome for genes that were probably being expressed at the time of extraction of the mRNA. Such a cDNA library is less complex than the genomic library and allows direct comparisons to be made between species, isolates or strains. An advantage of this approach is that the cDNA is present in the form of genes or parts thereof, free of introns and other extraneous DNA, whereas genomic DNA digests consist of restriction fragments that do not necessarily coincide with genes. In addition, differences in mRNA represent differences in biochemical activity of the cells of the organisms in question, which are more likely to be relevant in understanding the biochemical bases of differences between species, strains or isolates. A limitation of this approach may be that different genes are being expressed at different stages of the life cycle, or under different physiological or culture conditions; genes responsible for a given phenotypic difference may not be expressed at the time of mRNA extraction. Where possible, it is essential to control for such effects.

(v) Extraction of mRNA as above, incorporation into cell-free translation systems to produce proteins corresponding to the genes being expressed at the time of extraction. Such translation products would include antigens and other proteins of interest and, provided appropriate controls are used, may themselves be used to distinguish species, strains or isolates.

(vi) Identification and cloning of unique nuclear or mitochondrial DNA fragments, discovered by the above techniques, that either belong to, or are closely associated with specific genes, or are restricted to a particular species, isolate or strain, that can be used to probe the nuclear or mitochondrial genome for identification purposes. Such a sensitive probe is important for diagnostic applications. Specific genes may be searched out using probes developed for different organisms (e.g.

humans or bacteria) because the structure of many proteins is conservative and there is considerable homology between the appropriate genes from widely separated taxonomic groups. A useful such example would be the tubulin genes in nematodes, which probably vary slightly between strains resistant or susceptible to benzimidazole anthelmintics. These genes could be found in nematode extracts using a tubulin probe from fungi, for example. Once located and cloned, they could be sequenced and assessed for differences between the benzimidazole-resistant strains. If specific differences were found, they could be used as a sensitive and rapid probe for detecting such resistant strains without the need for other forms of assay.

These general techniques are very sensitive and usually require only small amounts of material, though they are technically complex, require rigorous standardisation and are not generally suitable for field studies at present. A number of protozoan species have been investigated by these methods, but work on the parasitic helminths is in its infancy. The Protozoa have the advantage of being clonable, which simplifies genetic analysis, whereas this is not generally practical for helminths, except in certain larval stages such as the rediae of trematodes or protoscoleces of *Echinococcus* species. It is important to note that results or conclusions that are based on analysis of nuclear DNA alone, without taking account of expression of the DNA (as in mRNA studies), need not necessarily give the same answers in strain variation studies as might be expected from comparisons based on phenotype. This is because nuclear DNA analyses utilise the entire genome, which includes material that may not be expressed at all. Thus, diagnostic characters need not necessarily be relevant biochemically or biologically. By the same token, biochemical or physiological characters may not be useful diagnostically, as is evident in the following example.

Organisms of the *Trypanosoma brucei* group in Africa, which are responsible for trypanosomiasis in both man and domestic and wild animals, have been subdivided into three subspecies on epidemiological and clinical evidence according to their infectivity to man (*T. b. brucei* does not infect man), and the type of sleeping sickness caused (*T. b. gambiense* causes chronic sickness mainly in West Africa, whereas *T. b. rhodesiense* causes acute sickness in East Africa). *T. b. rhodesiense* and *T. b. gambiense* both infect cattle and wild animals, which serve as reservoirs for human infection. The subspecies cannot be distinguished morphologically or immunologically (because of antigenic variation) and in the past unknown isolates were characterised either by infection of human volunteers or by infectivity to laboratory animals following incubation in human blood.

Analyses of enzyme polymorphism in these groups showed that *T. b. gambiense* could be distinguished from the other two by specific enzyme markers (see Gibson, Marshall and Godfrey 1980; Tait, Babiker and LeRay 1984) and could be classified as a separate subspecies, but *T. b. brucei* and *T. b. rhodesiense* could not be reliably separated by this method. Tait, Barry, Wink, Sanderson and Crowe (1985) examined this question further and found no *T. b. rhodesiense*-specific enzyme variants in 11 stocks. They concluded that *T. b. rhodesiense* is a subset of *T. b. brucei* that is able to infect man, that the two are not distinct subspecies, and hypothesised that the ability to infect man may lie in a particular combination of alleles or set of genotypes within the *T. b. brucei* genome. Thus, a reliable enzyme method for distinguishing the *rhodesiense* variant is not available. Nonetheless, for epidemiological reasons it is important to be able to recognise this variant.

Restriction enzyme analysis of the kinetoplast maxicircle DNA (equivalent to mitochondrial DNA from other organisms) from these three groups showed that they are so closely related that they could not be distinguished by this method (Borst, Fase-Fowler and Gibson 1981). Thus, this evidence brings into question the status of the *gambiense* variant as well. The stocks examined differed from each other at only a small number of polymorphic sites that did not correlate with any particular group; the range of differences was shown to be 0.01-0.03 substitutions per base pair, which, in other organisms, is in the range for regional differences in mitochondrial DNA (mtDNA) from the same species. More recent analysis of additional stocks (Gibson, Borst and Fase-Fowler 1985) identified two distinct maxi-circle subtypes that differed from each other and the other groups by 0.06-0.08 substitutions per base pair. These did not correlate with human infectivity but the findings show that subgroups can be identified by this technique. A recent study suggests that selection (or induction?) of human-infecting variants of *T. b. brucei* can take place in the tsetse fly: laboratory feeding of flies infected with a clone of *T. b. brucei* on uninfected human serum led to the appearance of variants that were resistant to human serum and therefore potentially infective to man (Rickman, Ernest, Dukes and Maudlin 1984).

Since mtDNA is normally inherited maternally (in organisms where the gametes differ substantially in size) and since it is believed to code for proteins restricted to mitochondria (with associated strict selection pressure for functional mitochondria), comparisons of mtDNA are not necessarily a good guide to variation in nuclear DNA. Trypanosomes contain a single mitochondrion which would be distributed in the same manner as the nucleus during cell division. But as sexual recombination has now been shown to occur during the life cycle, the nuclear and mitochondrial genomes need not necessarily be inherited together.

Therefore, it may be more fruitful to characterise or probe the nuclear DNA of these organisms to search for genes or specific base sequences that are associated with the ability to infect man in the *rhodesiense* variant. Two general approaches are possible: (i) preparation of mRNA, for a cDNA probe, from organisms (in tsetse flies fed on human serum) that might be expressing the appropriate genes or, (ii) a genomic screen of organisms from human or animal infections using appropriate selected digestion fragments as hybridisation probes.

cDNA sequences prepared from mRNA coding for the variable surface glycoprotein (VSG) genes were used as probes by Massamba and Williams (1984) to distinguish *T. b. brucei* and *T. b. rhodesiense* from *T. congolense, T. evansi, T. b. gambiense* and *T. vivax*, but *T. b. brucei* and *T. b. rhodesiense* could not be distinguished from each other. Additional hybridisation probes prepared from endonuclease digests of nuclear DNA from *T. b. brucei* did not hybridise with nuclear DNA prepared from any of the above species except *T. b. rhodesiense*, showing that the latter is closely related to *T. b. brucei*. Specific hybridisation fragments from *T. b. brucei* were used to distinguish different stocks of this species. Thus, this method has potential for screening isolates if probes can be developed that are species- or variant-specific. If specific *T. b. rhodesiense* probes can be developed, a test will then be available for parasites that have the potential to infect humans. These techniques could also shed some light on the specific biochemical changes in *T. b. rhodesiense* that are responsible for the ability to infect man.

Interesting insight into the interaction between hosts and parasite populations can be gained from DNA studies of isolates that differ in virulence. Virulent geographical isolates of *Babesia bovis* can be rendered avirulent by syringe passage through splenectomised calves; this can be reversed by rapid passage of the isolate through normal cattle. These strains can be differentiated antigenically. After many passages through splenectomised calves, the isolate loses first virulence and, later, infectivity to the tick vector. The molecular events controlling loss of virulence and infectivity were investigated by Cowman, Timms and Kemp (1984), who prepared cDNA clones from an avirulent isolate and cDNA probes from mRNA from both the parent virulent isolates and other isolates, both virulent and avirulent. They found a small number of genes that were transcribed mainly in avirulent parasites, but were weakly represented in other isolates also. Using these specific probes on genomic DNA digests they demonstrated polymorphism of these genes between different geographical isolates; this permits different isolates to be identified. Passaging an avirulent isolate through the tick vector changed the hybridisation pattern from that of the avirulent to one more closely resembling the virulent form of the same original isolate. Cowman *et al.* hypothesised that the process of attenuation selects minor avirulent subpopulations within an isolate that are expressing particular genes represented in the cDNA probes. The function of these genes is unknown and they need not necessarily be directly involved in the attenuation process. When an avirulent isolate is passaged through the vector, rapid selection takes place for the minor virulent subpopulations that are not expressing these genes. Cloning studies will clarify this question.

Recombinant DNA techniques are being employed to identify and characterise species and strains of many other protozoan parasites, including *Trypanosoma cruzi, Leishmania* spp., *Plasmodium* spp. and other groups of medical and veterinary importance. Progress in such work with helminths has been slower but the results are promising. Analysis of restriction fragments from digests of genomic DNA has been shown to be useful for identifying species of nematodes

(Curran, Baillie and Webster 1985). The bands corresponding to repetitive DNA fragments were specific for each species, but the method did not distinguish between strains of the same species, which is important in many applications. Therefore, it will be necessary to analyse mtDNA, or genomic DNA, by recombinant methods to make progress in this area.

At present the functional significance of the polymorphisms and other genomic variations detected by these techniques is not at all clear. Insufficient information is available for different groups to use nucleic acid studies alone to classify strains, subspecies or species; in conjunction with other techniques, however, DNA work is especially valuable for diagnosis. Our understanding of host-parasite relations and interactions will be significantly improved by these studies. Once important genes have been identified, it will be possible to monitor their expression during the life cycle and perhaps to understand the significance of the polymorphisms that appear to be common for some genera, though perhaps not all.

6.4.3 *Metabolic studies*

Phenotypic diversity at the functional level can be detected within and between species by investigation of the metabolism of organisms. This type of variation is detected in metabolic pathways or their regulatory processes, in susceptibility to antiparasitic drugs, in variable development times, or as antigenic variation. In this section we will discuss principally the first type of variation, since it has interesting implications for the study of the host-parasite relationship. Variations in energy metabolism, in particular, are especially interesting, because it has been generally believed that these pathways, being essential to an organism's metabolic economy, would tend to be conservative because of strong selection pressure for efficient energy production. Since energy metabolism does in fact vary, we need to reconsider our views and perhaps reassess the roles both of host influences on such pathways and stochastic events in the evolution of parasites. A brief survey of metabolic variation in some species of parasites will illustrate the types and extent of such variation as we know it at present.

The carbohydrate metabolism of *Echinococcus granulosus* has been shown to vary with the nature of the host and with different geographical locations. Variation occurs in both the larval stage (protoscolex) and in the adults. The proportions of excreted end products of aerobic carbohydrate metabolism in this species from different host-parasite systems are summarised in Table 6.1. All the isolates tested are slightly different, but in some cases the variation is extreme: for example, adults from the wallaby/dingo cycle in Queensland produce almost no detectable succinic acid, whereas adults from Kenyan cycles produce up to 50%.

Cross-infection studies and genetic analyses are required to determine whether these are direct genetic effects resulting from either isolation processes or host selection of the parasites, or whether they are due to physiological influences of different hosts on the parasites. Those parasites producing higher proportions of lactic acid or ethanol would gain significantly less ATP from the glucose units consumed. Yet, protoscoleces from the horse strain, producing 59% lactic acid aerobically, consumed considerably less glycogen when incubated *in vitro* than the UK sheep strain, which produced 22% lactic acid (McManus and Smyth 1978). At the same time, the O_2 consumption was higher in the sheep strain. Does this mean that ATP requirements are higher for the sheep strain? If so, is this evident physiologically in, for example, a higher rate of growth? Perhaps the sheep presents more immunological challenges for the sheep strain to combat, thus

Table 6.1: *Echinococcus granulosus* intraspecific variation in end products of carbohydrate metabolism excreted during aerobic incubation *in vitro*

Protoscoleces: Origin	% of Total Excreted End Products:			
	Acetic acid	Lactic acid	Succinic acid	Ethanol
Sheep – Kenya*	38	23	38	–
Sheep – UK*	55	17	28	–
Sheep – UK**	58	22	14	7
Sheep– Tasmania+	37	21	11	32 (n=1)
Cattle – Kenya*	35	10	55	–
Goat – Kenya*	48	24	28	–
Horse – UK **	25	59	8	8
Camel – Kenya*	13	38	48	–
Man – Kenya*	25	45	30	–
Adults: Origin				
Natural/dog – Kenya*	18	31	51	–
Man/dog – Kenya* (experimental)	33	16	51	–
Sheep/dog – NSW+	23	21	36	20 (n=1)
(isolates from different	26	17	42	15 (n=1)
locations)	21	27	21	31 (n=4)
Sheep/dog – Tasmania+	31	26	9	34 (n=2)
Wallaby/dingo – Queensland+	21	36	0.4	43 (n=1)

– not determined
* calculated from McManus 1981
** calculated from McManus and Smyth 1978
+ Behm, Bryant and Thompson unpublished

dictating a higher ATP consumption. Alternatively, physiological conditions in the sheep may be such that a higher rate of ATP production is possible; such favourable conditions may not occur in the horse. Similar arguments may also be presented for the Kenyan human isolate. Clearly, more samples need to be taken, and more detailed investigations made, so that we can decide between the plethora of possible explanations.

Early studies of the energy metabolism of *Hymenolepis diminuta in vitro* by Read (1956) and Laurie (1957) showed that a major end product of carbohydrate metabolism was lactic acid. Read determined that 71-91% of total acid excretion was lactic acid, whereas Laurie found a greater variation, 37-98%. Later, Fairbairn's group assayed the other metabolic end products from this parasite, succinic and acetic acids, and showed lactic acid to constitute only from 5 to 40% of the total excreted acids (Fairbairn, Wertheim, Harpur and Schiller 1961; Watts and Fairbairn 1974). Younger (6 day) worms were shown to excrete a greater proportion of lactic acid than mature (14 day) worms, and succinic acid was the major end product in the older worms. Coles and Simpkin (1977) dissected mature (14-21 day) worms into quarters along their length and showed that the anterior 2 cm of the worms produced only 4% succinic acid compared with 36% in the remaining three quarters of the worms. In each case the major end product was acetic acid. More recently, Ovington and Bryant (1981) demonstrated that this parasite, in their system, produced around 84% lactic acid, with succinic acid hardly detectable at 4%.

Clearly, *H. diminuta* in all the examples quoted above has the capacity to produce each of the three end products of carbohydrate metabolism, but the proportions in which they are excreted vary considerably and differ not only between worms of different ages and between different parts of the adult worm, but also between different laboratories around the world. Are the geographical differences true 'strain' differences or do the worms merely reflect in their end products the different environmental conditions under which they are reared and incubated? In the examples cited above, different rat strains were used in different laboratories, the rat sex was not specified, the intermediate host was not stated in most cases, and the rats were infected with different numbers of cysticercoids. It is well established that growth rate and egg production in *H. diminuta* is influenced by rat strain, infection density and diet (see, for example, Boddington and Mettrick 1981). All these factors – and many more – may affect the metabolic development of *H. diminuta*, both by genetic selection (in the rat or intermediate host) and by physiological adaptation. In addition to these variables, the initial isolation event (acquisition of *H. diminuta* by a particular laboratory) would determine the genetic limits either upon which selection could act or within which physiological adaptation could take place.

Some recent experimental work has attempted to sort out some of the variables discussed above. An interchange was made of *H. diminuta* cultivars between the Department of Zoology, University of Toronto (where the parasite, designated

'UT', had been grown for many years using the flour beetle, *Tribolium confusum*, as intermediate host) and the Department of Zoology, Australian National University (where the parasite, designated 'ANU', had been grown for many years using the meal worm *Tenebrio molitor* as intermediate host). Comparative biochemical studies with both 'strains' were then performed in each laboratory using Wistar rats and both species of intermediate host. Some preliminary results are summarised in Table 6.2.

Table 6.2: *Hymenolepis diminuta*: acidic end products excreted during aerobic incubation *in vitro* by two worm 'strains' in different laboratories

Strain	Intermediate host	Succinic	Acetic (% of Total)	Lactic	n	Reference
UT	*Tr. c.*	28	31	41	5 × 5-6 worms	1
UT	*Te. m.*	35	38	27	5 × 5-6 worms	1
ANU	*Tr. c.*	23	20	57	5 × 5-6 worms	1
ANU	*Te. m.*	27	31	42	5 × 5-6 worms	1
UT	*Tr. c.*	39	58	3	3 × 1 g wet wt	2
ANU	*Tr. c.*	23	9	58	3 × 1 g wet wt	2
UT	*Tr. c.*	41	34	24	5 worms	3[+]
UT	*Tr. c.*	44	40	16	5 worms	3[+]
UT*	*Tr. c.*	30	34	36	5 worms	3[+]
UT*	*Tr. c.*	41	34	25	5 worms	3[+]
ANU	*Tr. c.*	47	30	23	5 worms	3[+]
ANU	*Tr. c.*	5	9	86	5 worms	3[+]
ANU*	*Tr. c.*	24	21	56	5 worms	3[+]
ANU*	*Tr. c.*	47	42	11	5 worms	3[+]

Tr. c. = *Tribolium confusum*, *Te. m.* = *Tenebrio molitor*
[+] These figures represent the mean of the incubation products of 5 worms, individually incubated, from a single rat in each case.
* Rat fasted for 24 h before collection of worms.
1. Mettrick and Rahman 1984
2. Kohlhagen, Behm and Bryant 1985
3. Behm, unpublished

The first study, in Toronto, showed that the two 'strains', when passaged through the same intermediate host species, can be distinguished by their end products, the major difference being that the ANU strain produced more lactic and less succinic acid than the UT strain. But, the proportions of these end products overlapped when the worms were passaged in different intermediate host species. Differences were also recorded in the steady-state concentrations of some glycolytic intermediates and adenine nucleotides. Thus, the intermediate host species influences the subsequent *in vitro* metabolism of the adult worms. In the second study, in Canberra, the same general differences were evident between the 'strains', but the differentiation was more extreme, in that the UT 'strain' produced almost no lactic acid, and the ANU 'strain' very little acetic acid. Clearly, other factors also influence end products of *H. diminuta*; these could include differences in the intermediate host populations, rat variations, differences in diet and many other variables. This work has been extended to survey individual worms in individual rats (study 3) which has led to some interesting observations. In this trial, all the worms from a given individual rat (originally infected with 15 cysticercoids) showed the same proportions of end products, but worm populations from different rats (in otherwise apparently identical infections) differed from each other. In some cases the variation was minor, in others it was extreme. Clearly the host rat has an important effect on the metabolism of the adult worm which is superimposed on any strain difference between worm populations. Further studies are in progress to elucidate this problem.

A comparative study of the activities of enzymes of carbohydrate metabolism in the two 'strains' of *H. diminuta* (part of study 2, above) showed consistent and significant differences between the two 'strains' in important enzymes, particularly pyruvate kinase, lactate dehydrogenase, malic enzyme, fumarase and fumarate reductase, which account for the differences in excreted end products. A more detailed investigation is now required on enzyme activities in worms from different rats.

Morphological differences and differences in activities of brush border enzymes have also been demonstrated between *Hymenolepis* 'strains' from different laboratories cultured under identical conditions (Pappas and Leiby 1986a,b). This work showed that there is considerable morphological variability in this species, both within and between 'strains', but only a small selection of variables could be used to characterise a given 'strain'.

Thus, although apparent 'strain' differences are evident in *H. diminuta* from different laboratories around the world, they are difficult to characterise by metabolic end product excretion which is influenced by many factors in the worms' environment. Other criteria, particularly genetic studies, are required to confirm the existence or otherwise of stabilised 'strains' in this species. How these 'strains' relate to wild populations of this parasite could form the basis of an interesting evolutionary study.

Variations of energy metabolism in *Haemonchus contortus* have been discovered during the course of investigations into the metabolism of strains of this parasite resistant to the benzimidazole anthelmintics. *H. contortus* adults utilise both aerobic and anaerobic pathways of energy metabolism, producing as major end products acetic and propionic acids, plus n-propanol and ethanol. Under aerobic conditions *in vitro* they also produce significant quantities of CO_2.

The fumarate reductase complex in this parasite has been proposed as a target for thiabendazole and other benzimidazole anthelmintics. The activity of this enzyme has been examined in a variety of benzimidazole-susceptible or -resistant strains of *H. contortus* held in different laboratories. The results of these determinations were evaluated by Bryant and Bennet (1983) who concluded that the absolute activity (as measured) of this enzyme varied between different isolates and that there was no necessary correlation between fumarate reductase activity and tolerance of these strains to the benzimidazole anthelmintics. Since fumarate reductase was thought to play a central role in energy metabolism, it was surprising to find such variation in activity; the role of both this enzyme and the associated anaerobic pathways in the metabolic economy of this species evidently requires more detailed study.

The glucose metabolism of two North American strains of *H. contortus* was investigated by Rew, Smith and Colglazier (1982), who showed that, *in vitro*, a cambendazole-resistant strain excreted end products in the same proportions as a susceptible strain. There was no difference between the strains in $^{14}CO_2$ production from ^{14}C-labelled glucose, nor in glucose transport. But when thiabendazole was included in the incubation medium, significant differences were seen in the *in vitro* end products. These are presented in Table 6.3. Thus, differences were not evident between the strains until they were tested in the presence of an anthelmintic compound.

Table 6.3: *Haemonchus contortus:* metabolic end products excreted *in vitro* by adults of two strains from the USA*

	Susc	Susc + 5mM TBZ	CBZR	CBZR + 5mM TBZ
Acetic acid	0.17	0.13	0.25	0.31
Propionic acid	0.07	0.08	0.16	0.16
n-Propanol	0.41	0.21	0.27	0.43
Ethanol	0.17	0.07	0.07	0.23
Total	0.84	0.49	0.75	1.13

Susc = susceptible; CBZR = cambendazole-resistant; TBZ = thiabendazole
* from Rew, Smith and Colglazier 1982; all units are mmol mg protein^{-1} h^{-1}.

In contrast, three Australian strains of *H. contortus* showed differences without *in vitro* drug treatment. Bennet and Bryant (1984) compared laboratory-maintained benzimidazole-susceptible, thiabendazole-resistant and mebendazole-resistant strains. A summary of some of their results is given in Table 6.4. The mebendazole-resistant strain differed from the other two by excreting smaller quantities of propionic acid and greater amounts of ethanol. But fumarate reductase activity in this strain was similar to that in the susceptible strain, whereas it was lower in the thiabendazole-resistant strain. At the same time, anaerobic $^{14}CO_2$ production from U-^{14}C-glucose was highest in the thiabendazole-resistant strain yet, aerobically, the mebendazole-resistant strain had the highest rate of production.

Table 6.4: *Haemonchus contortus*: comparison of *in vitro* energy metabolism in adults of three strains from Australia

	Susceptible	TBZ-resistant	MBZ-resistant
1. Excreted Metabolic End Products (mmol/g wet wt)*			
Acetic acid	79.2 ± 62.3	84.0 ± 43.2	64.9 ± 28.4
Propionic acid	79.7 ± 41.3	83.1 ± 20.0	49.2 ± 24.0
n-Propanol	44.2 ± 22.4	53.0 ± 18.2	49.5 ± 27.9
Ethanol	28.5 ± 27.4	17.8 ± 11.0	48.5 ± 48.0
2. Fumarate Reductase Activity (nmol/mg protein/min)[+]			
	12.1 ± 3.7	3.0 ± 1.1	11.7 ± 1.5
3. $^{14}CO_2$ Production from U-^{14}C-glucose (cpm x 10^{-3}/g)*			
Anaerobic	15.4 ± 2.3	26.8 ± 5.0	15.4 ± 4.6
Aerobic	39.5 ± 8.4	50.4 ± 21.5	77.6 ± 21.5

TBZ = thiabendazole
MBZ = mebendazole
* Bennet and Bryant 1984
[+] Bryant and Bennet 1983

Clearly, these variations do not account directly for the resistance of the strains to the benzimidazole anthelmintics because the thiabendazole-resistant strain is cross-resistant to mebendazole and vice-versa. We cannot therefore decide from this information whether these variations are *necessarily* linked to the resistant phenotypes or whether they are chance isolation events. It is possible, of course, that resistance to similar compounds may manifest itself at the molecular level by more than one mechanism.

The differences that have been observed in energy metabolism in these parasites imply different rates of ATP synthesis. We cannot assess this quantitatively at this stage because the precise pathways of synthesis of the end products have not yet been conclusively demonstrated. But, despite their differences in energy metabolism, these three strains of *H. contortus* survive and reproduce effectively in the laboratory, though their rates of maturation, egg production and pathogenicity in the sheep are different and characteristic for each strain. Are any of these strains more 'fit' than the others? This is almost impossible to assess, since 'fitness' is difficult to define in any situation. We do not know what happens to the metabolism of the resistant strains once selection pressure (in the form of anthelmintic treatment at every generation) has been removed. The result of withdrawal of treatment could be an important indicator of any role the treatment may have in influencing the types of energy metabolism in these parasites. Similarly, we do not know how well these strains would survive in the field: this is an important criterion for assessing 'overall' fitness, which would be quite different from fitness in the laboratory environment. Some answers to these questions may be obtained by examining field strains that have not been exposed to laboratory selection.

Trichomonas vaginalis and *Tritrichomonas foetus* are parasitic Protozoa that have an unusual anaerobic energy metabolism that is fundamental to the antiprotozoal activity of the 5-nitroimidazole compounds such as metronidazole. The parent compounds do not kill trichomonads, but when they are enzymatically reduced in the hydrogenosomes to unstable nitro radical metabolites, they are highly cytotoxic. Since the introduction of these compounds in clinical medicine several decades ago, strains of *T. vaginalis* have appeared that are resistant to this treatment. Metabolic activation of the 5-nitroimidazoles is closely linked to energy metabolism in these parasites, because the responsible reducing activity is associated with the hydrogenosomal pyruvate:ferredoxin oxidoreductase complex, which plays a central role in glucose catabolism. Thus, pyruvate is the electron donor for both energy metabolism and metronidazole reduction. It is instructive, therefore, to compare energy metabolism in strains with different susceptibilities.

Resistant and susceptible strains of *T. vaginalis* can be distinguished *in vitro* by their different rates of uptake of ^{14}C-metronidazole under aerobic conditions (Müller and Gorrell 1983). Resistant strains take up the drug more slowly than the susceptible strains. Anaerobically, however, it is not possible to make this distinction. The rate of uptake of metronidazole *in vitro* is a measure of the rate of intrahydrogenosomal metabolism of this compound, because its entry into the parasites is passive and dictated by relative internal and external concentrations. Metabolism thus effectively lowers the internal concentration, causing a greater rate of influx. The presence of oxygen significantly inhibits the metabolism of metronidazole, but this occurs to a greater extent in resistant strains than in susceptible ones. Thus, resistance to metronidazole in these strains is in some way related to the ability of the hydrogenosome to activate it in the presence of oxygen.

Table 6.5: *Trichomonas vaginalis* : energy metabolism *in vitro* of different isolates resistant or susceptible to metronidazole*

(1) Anaerobic Metabolism

	H_2		Acetic		Lactic		$\dfrac{\text{Lactic}}{\text{Acetic}}$
Susceptible							
ATCC 30001[+]	12.2	± 1.2	14.2	± 2.6	47.2	± 4.2	3.3
NYH 209	18.0	± 8.5	13.9	± 11.8	14.5	± 4.3	1.0
NYH 2721	12.2	± 6.9	8.9	± 5.0	18.6	± 1.5	2.1
NYH 286	25.0	± 2.8	16.3	± 1.9	29.2	± 8.9	1.8
Wien A	13.7	± 10.3	9.8	± 5.2	38.0	± 15.4	3.9
Resistant							
IR-78	13.3	± 11.0	9.2	± 4.5	30.8	± 5.4	3.3
Fall River	43.3	± 17.1	16.3	± 3.7	26.3	± 0.5	1.6
Boston	11.5	± 4.9	6.9	± 0.7	22.1	± 14.1	3.2

(2) Aerobic Metabolism

	Oxygen Uptake
Susceptible	
ATCC 30001[+]	24.2 ± 1.5
NYH 209	34.4 ± 9.6
NYH 272	39.9 ± 2.3
NYH 286	52.5 ± 9.8
Wien A	53.5 ± 7.3
Resistant	
IR-78	37.0 ± 4.3
Fall River	41.4 ± 10.1
Boston	36.4 ± 10.0

*Data from Müller and Gorrell 1983. Parasites incubated in the presence of 50 mM glucose. All units are nmol min^{-1} mg^{-1} protein.
[+]A laboratory strain not infective to mice.

This may indicate a change in the pyruvate:ferredoxin oxidoreductase complex or its local environment. It is not clear from *in vitro* studies how physiologically relevant these observations are, because the oxygen concentrations *in vivo* are not known but are unlikely to be either zero (completely anaerobic) or 20% (air saturation). Possibly a low *in vivo* oxygen concentration is sufficient to affect metronidazole metabolism in resistant and susceptible strains, or resistant strains may be more aerotolerant than susceptible ones.

Müller and Gorrell (1983) surveyed the *in vitro* energy metabolism of isolates of *T. vaginalis* resistant and susceptible to metronidazole. Some of their results are summarised in Table 6.5. Aerobic incubation had no effect on acetate or lactate production (not shown). Hydrogen and acetic acid originate from the hydrogenosomes, whereas lactic acid production is cytosolic. There is considerable variation between isolates in their metabolic end products,. particularly in the relative contribution of hydrogenosomal and cytosolic metabolism as shown by the lactic/acetic ratios. But there is no consistent difference in end products or oxygen uptake between these metronidazole-susceptible or -resistant strains. Müller and Gorrell also assayed the activities of the enzymes pyruvate:ferredoxin oxidoreductase, NADH oxidase and NADPH oxidase but found that, although activities varied significantly between isolates, there was no correlation with susceptibility. Therefore, in these examples, resistance to metronidazole is not accompanied by any gross changes in energy metabolism, but is presumably due to additional (at present unknown) barriers to the transfer of electrons to metronidazole that become detectable only when oxygen is present to compete as an electron acceptor.

The synthesis of acetate in the hydrogenosomes of trichomonads may be an energy-generating process. Thus, hydrogenosomal acetate production may generate more ATP per unit of glucose than cytosolic lactate production. If this is so, some of the *T. vaginalis* strains that produce higher quantities of lactic acid may have an energetic disadvantage. There is no evidence for this, however, and such variation as is seen *in vitro* is clearly compatible with effective growth and reproduction, at least in the laboratory/mouse environment. Strains recently isolated from the field need to be examined to clarify this question.

Tritrichomonas foetus is metabolically slightly different from *T. vaginalis*. Its metabolic end products are hydrogen, acetic and succinic acids, and ethanol. It is sensitive to metronidazole, and resistant strains have been identified in the field. These field strains have not been characterised in detail, but it appears that, as with *T. vaginalis*, their resistance is aerobic only. Stable resistant strains have also been developed by laboratory selection and cloning *in vitro* (Kulda, Cerkasov and Demeš 1984). These differ from the field strains in that resistance *in vitro* can be detected anaerobically.

T. foetus laboratory strains with anaerobic resistance have been shown to have a highly modified energy metabolism. The activities of pyruvate:ferredoxin oxidoreductase and hydrogenase are significantly diminished or absent, and acidic end products and hydrogen are not produced. Instead, pyruvate is metabolised by cytosolic reactions to ethanol which is the major end product (Cerkasovová, Cerkasov and Kulda 1984). At the same time, the rate of glycolysis is significantly increased. But the growth rate and size of these resistant strains is considerably diminished; presumably, less ATP is produced because the phosphorylations associated with acetate production do not occur. This evidence demonstrates the important role of the hydrogenosomal reactions in the metabolic economy of these parasites.

Clearly, resistance to metronidazole in these strains of *T. foetus* is brought about by the loss of enzyme activities responsible for the metabolism of the drug. This loss was a property of all the resistant strains derived *in vitro* from the parent KV1 strain used by Kulda *et al.* (1984), and the resistance remained stable, in the laboratory, when drug selection pressure was removed. But the metabolic cost of this adaptation is considerable, and it is not clear whether these resistant strains can infect laboratory hosts or cattle. It seems likely that they would be unable to compete successfully *in vivo*, especially in the field where the selection pressures would be more stringent and variable than in the laboratory. Examination of field-derived resistant strains of this parasite may answer this question. If the anaerobic type of resistance is indeed found in the field, and this correlates with loss of the hydrogenosomal reactions discussed above, then it will be evident that this organism can survive and multiply without the additional energy-generating reactions. Thus, although *T. foetus* obviously has the capacity to survive *in vitro* without the hydrogenosomal reactions, there is no evidence at present that this actually occurs in the field.

6.5 Resistance to antiparasitic compounds

The widespread use of antiparasitic compounds in clinical, veterinary and agricultural practice has led to the development of resistance to some compounds in certain parasite populations. Resistance has also been induced experimentally in laboratory populations of parasites, and it presents a serious and growing problem in the treatment or prophylaxis of parasitic infections. The dangerous resurgence of malaria all over the world is due, in part, to a combination of (i) increasing resistance of the mosquito vectors to insecticides, and (ii) the appearance and spread of populations of the parasites resistant to chloroquine and other compounds used for prophylaxis or treatment (see Peters 1985).

The development of resistance by parasite populations is an almost inevitable outcome of the evolutionary processes set in train by the heavy selection pressure of drug treatment. For both theoretical and practical reasons it is important to understand the biochemical and genetic mechanisms that enable organisms and populations to become resistant to antiparasitic drugs. Then we can use new compounds more rationally, in ways that reduce the probability of inducing resistance. We can also overcome or circumvent the problem by modifying existing compounds or by using combinations of different compounds with complementary or synergistic effects.

6.5.1 *Genetics of resistance*

Inheritable drug resistance in an individual or in a population may be due to the activity of either a particular allele of a single gene, or a combination of alleles of different genes ('polygenic') (see McKenzie 1985). The selection pressure of the drug treatment on the population results in an increase in the frequency of the alleles conferring resistance in the treated population. In parasites, this change in frequency occurs because, in routine field use, it is rare for a compound to kill or remove all individuals of a parasite sub-population within a host. This is because not all individuals are exposed to an effective dose, for a number of reasons, including:

(i) the distribution of the compound within the parasite's environment is not uniform, so that some individuals, who may be in slightly different locations, are exposed to a lower drug concentration.

(ii) some stages of the life cycle within the host are more sensitive to the drug than others, and some individuals may have a different rate of development from others, or

(iii) some individuals possess rare alleles of certain genes that permit them directly to survive a drug concentration that removes a large proportion of the population (i.e. genes conferring drug resistance).

Selection by drug pressure on the parasite sub-population probably occurs most frequently by mechanism (iii), though there may also be genetic components to mechanisms (i) and (ii).

The survivors of drug exposure contribute their genes to the next generation. If the drug selection pressure is maintained, the frequency of those alleles conferring resistance to the compound will increase in the population, sometimes very rapidly, at a rate determined by (i) the difference in fitness (i.e. the numbers surviving to contribute genes to the next generation), (ii) the intensity of the selection pressure (i.e. the dose rate and frequency of dosing), and (iii) the proportion of the total interbreeding population that is exposed to the drug at each generation. Whether the alleles conferring resistance subsequently persist at the higher frequency in a population depends to a large extent on the relative overall fitness of the resistant phenotype in the presence or absence of drug selection pressure.

Resistance conferred by a single gene usually occurs when a population is exposed to high and persistent concentrations of the toxicant that remove almost the entire population. A rare mutation may then be selected that confers on the initial heterozygote a level of resistance *outside and above* the normal variation in the population (see Fig 6.2). This appears to be the case in resistant populations of most arthropod pests. If, as a side effect, a compound also increases the rate of mutation, the probability of this type of resistance developing in a population may increase. The large difference in fitness between resistant and susceptible phenotypes in a single gene system means that, once the mutation appears in a population, resistance will develop rapidly provided the selection pressure remains high. Experimentally, the difference between related resistant and susceptible populations is seen as allelic frequency differences at a single, specific locus.

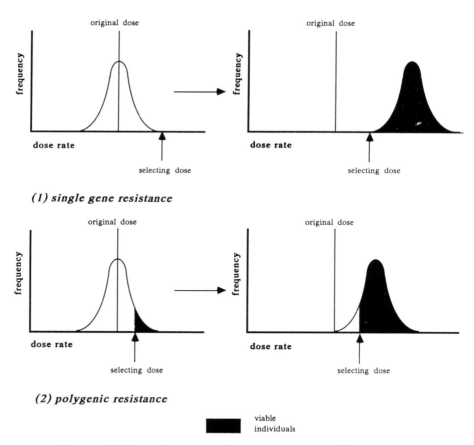

Figure 6.2 Development of Drug Resistance (based on McKenzie 1985)

Resistance that is polygenic is more likely to occur when the selection removes a smaller proportion of the population, and where the toxicant is not persistent in the organism's environment. In this case, selection acts initially on phenotypes *within* the existing population variation. As many biochemical, physiological, developmental or environmental factors within this variation may contribute to an individual's tolerance of the compound, a variety of genes will be implicated. This type of resistance usually develops more slowly than single gene resistance because, first, a number of rare and separate genes is being selected simultaneously and second, it does not include very large changes in fitness at each selection cycle. It also requires a sufficiently large population to provide the variation in susceptibility upon which selection can act. When isolated populations are being challenged by the same drug, they may resolve their problem in different ways, depending on their existing phenotypic variation and the nature of the selection pressure. This means that separate populations of the same species may produce different resistant genotypes. 'Separate' in this context means that the populations do not interbreed due to host or geographic isolation. At the molecular level, therefore, different populations may overcome the toxicity of a drug by different mechanisms. Experimentally, for polygenic resistance, differences between resistant and susceptible populations are seen as allelic frequency differences at a variety of loci. For *closely related* populations the observed differences are probably directly related to resistance. For separate and less closely related parasite populations, however, the allelic frequency changes due to resistance are superimposed on *existing* allelic frequency differences between the original, susceptible parent populations. Therefore, a large number of allelic differences may be detected between susceptible and resistant parasite populations, *only some of which are directly implicated in resistance.* Similarly, at the biochemical level, the precise changes due to resistance are superimposed on any pre-existing biochemical differences between the populations in question. In addition, if the initial selection cycles remove a very large proportion of the original population, and if there is no significant recruitment of genes from outside the population, then components of random genetic drift may also enter the scene, even for closely related populations. This gives rise to genetic and biochemical changes that have no direct, or causal, relationship with resistance mechanisms. These factors seriously complicate our attempts to elucidate the molecular mechanisms of polygenic resistance, particularly when the precise mode of action of a compound has not been determined.

The basic genetic systems in parasites have not been investigated in detail, and there may be unusual features not common in other groups of animals. For example, gene amplification by the division of chromosomes (endomitosis) occurs in somatic cells during development in strongylid nematodes, and mitochondrial (kinetoplast) DNA (which is cytoplasmically inherited) plays an important role in kinetoplastids. Genetic studies of resistance in parasites are not well developed, but they could greatly assist in unravelling the molecular mechanisms of resistance (see

Le Jambre 1985). Resistance is apparently polygenic in some nematode populations, where there is some evidence for maternal inheritance of resistance, whereas single gene resistance appears to be more common in protozoan systems. In some parasite/drug combinations, resistance has developed after a single treatment. For example, hycanthone resistance in *S. mansoni* developed after a single exposure; this was probably due to selective destruction of gametes in the females (Jansma, Rogers, Lin and Bueding 1977).

6.5.2 *Molecular mechanisms of resistance*

It is not possible with limited space to do justice to the enormous variety of resistance mechanisms that undoubtedly exist – most undescribed – in parasites. In this section we shall list and discuss the theoretical possibilities, with selected examples where these can be found. It is quite probable that parasites employ a combination of these mechanisms in developing resistance, especially where it is polygenic.

(a) *Loss of specific uptake mechanisms for a compound*
To be effective, a drug must first enter or bind to the target parasite. The entry of a compound by diffusion may not permit the accumulation of a sufficiently high concentration within the parasite to be toxic. Some compounds are effective only because the parasite has specific (usually active) transport mechanisms. Loss, by genetic deletion, of the transport system would diminish the activity of the drug. Provided the normal function of the transport system concerned could be compensated for, resistant parasites would result.

Red blood cells infected with *P. berghei* possess high-affinity (ATP-dependent) and low-affinity binding sites for chloroquine, but the uptake rates for the drug are different in cells infected with susceptible and resistant strains (Diribe and Warhurst 1985). In cells housing the resistant strain 'RC', uptake is not energy-dependent but occurs more slowly via the energy-independent, low-affinity sites. The ATP-dependence of the high affinity site may be related to a requirement for proton pumping into the digestive vacuoles, where haemoglobin is digested (see Section 5.3.9). A similar difference in the kinetics of chloroquine uptake is seen between red blood cells infected with resistant and susceptible strains of *P. falciparum* (Verdier, LeBras, Clavier, Hatin and Blayo 1985). Experimental resistance to Formycin B in *L. tropica* is correlated with greatly reduced transport of the compound into the cells, indicating a change in the membrane components responsible for uptake (Rainey and Santi 1984).

(b) *Loss of the molecular binding site or target*
If a parasite can survive the loss of the specific molecule to which the drug binds, then the drug will not be effective. In the case of chloroquine resistance in

Plasmodium, there is some evidence that the high-affinity binding site in plasmodial parasites is ferriprotoporphyrin IX, a toxic product of the degradation of haemoglobin (see section 5.3.9). Some chloroquine-resistant strains of *P. berghei* do not accumulate ferriprotoporphyrin IX. Instead, they export free haem, which is a product of modified mechanisms for degrading haemoglobin (Wood *et al.* 1984).

In a laboratory-induced mutant line of *L. donovani* that was selected with tubercidin (an analogue of adenosine – see Section 5.3.6), adenosine kinase activity is lacking. The mutant line was shown to incorporate adenosine at a much lower rate than the parental line (Iovannisci and Ullman 1984). The absence of adenosine kinase persisted when drug selection was removed, suggesting a genetic deletion, but it remains to be seen whether such a mutant, with less efficient purine salvage pathways, would be viable in the field.

(c) *Structural modification of the binding site*

A slight modification to the binding site of the target molecule – which may or may not affect its overall metabolic function – may diminish the affinity of the target for the drug, so that much of the drug's efficacy is lost. Resistance to the benzimidazole carbamates in strongylid nematodes (*H. contortus, T. colubriformis*) and in certain fungi (*Aspergillus nidulans*) is correlated with reduced binding of the compounds to the tubulin of resistant strains (Lacey 1985; Davidse and Flach 1977). In *P. chabaudi*, dihydrofolate reductases from pyrimethamine-sensitive and -resistant strains are structurally slightly different, such that pyrimethamine does not competitively inhibit the enzyme from the resistant parasites (Sirawaraporn and Yuthavong 1984).

(d) *Compensation for loss of activity of the target*

Where the activity of a target (e.g. an enzyme) is inhibited, compensation may be achieved by either synthesising a greater amount of the target molecule, or by utilising bypass pathways, where these are available. The genetic basis for these changes is usually by gene amplification, that is, the development of multiple copies of the gene coding for the affected molecule, so that larger amounts of the target molecule, or possibly the bypass enzymes, are synthesised.

In laboratory-selected lines of *L. tropica* that are susceptible or resistant to methotrexate (a dihydrofolate reductase inhibitor) or 5,8-dideaza-10-propargyl folate (a thymidylate synthetase inhibitor), the bifunctional enzyme is produced in higher concentrations by the resistant cells, by amplification of the appropriate DNA sequence (Washtien, Grumont and Santi 1985). The kinetic and structural properties of the enzyme are not altered. Similarly, a strain of *P. falciparum* resistant to pyrimethamine has a 30-80-fold increase in activity of the target enzyme, dihydrofolate reductase, with no change in kinetic properties (Kan and Siddiqui 1979).

Some strains of *P. falciparum* resistant to sulfadoxine (which inhibits dihydropteroate synthase, the first step in *de novo* synthesis of dihydrofolate; see

Section 5.3.8) require a 'dialysable serum factor' to grow *in vitro*. This factor may be folic acid, which is a normal constituent of human serum (Watkins, Sixsmith, Chulay and Spencer 1985). Thus, these parasites bypass the *de novo* pathway by salvaging folate from the host. This is an interesting example of parasite metabolism, under drug pressure, approaching more closely that of the host.

(e) Loss of the activation mechanism for a compound

This mechanism is a special case of (b) or possibly (c) above, where the molecular binding site in the parasite is in fact an enzyme or system that modifies the compound to its active derivative, which then attacks additional targets within the parasite.

The rate of reduction of metronidazole to nitro anion radicals (see Section 5.3.10) in trichomonads is diminished in resistant strains, by various mechanisms. Laboratory-selected strains of *T. foetus* that are resistant under *anaerobic* conditions have lost pyruvate:ferredoxin oxidoreductase activity (Cerkasovová *et al.* 1984). Since this seriously cripples the parasites' energy metabolism and slows its growth rate, it is unlikely to be a common mechanism of resistance occurring in the field.

Normal *in vivo* environmental conditions for these organisms probably include a low level of oxygen, and aerobic resistance has developed in some strains. In resistant strain 85 of *T. vaginalis*, several changes have been observed in the parasites' capacity to reduce metronidazole. Hydrogen normally inhibits metronidazole reduction under anaerobic conditions in susceptible strains, but in the resistant strain it is less effective (Lloyd and Kristensen 1985). This indicates a change in the electron-transferring pathway. At the same time, the resistant strain appears to be deficient in oxygen-scavenging mechanisms, because at low external oxygen concentrations there is sufficient intracellular oxygen to compete for the electrons and inhibit nitro radical production (Lloyd and Pedersen 1985). In the susceptible strain under the same conditions, however, there is essentially no intracellular oxygen because it has been removed by scavenging enzymes. Consequently, metronidazole reduction is not inhibited. Therefore, at least two mechanisms apparently combine to confer resistance in this strain.

(f) Increased protective mechanisms to minimise damage

If damage is more indiscriminate or binding due to a compound is less selective than in (d) above, for example where the active species are free radicals, an increase in enzymes protecting against the damage would interfere with the activity of the drug. Free radical scavenging enzymes, such as superoxide dismutase, catalase, and glutathione peroxidase/reductase fall into this category. An increase in their activities, possibly achieved by gene amplification, would diminish the free-radical-induced damage.

(g) *Improved detoxification mechanisms*
A parasite that is able to metabolise a compound rapidly and efficiently to less toxic metabolites would be less susceptible to that compound. Most helminths are able to detoxify compounds by reduction or hydrolysis, followed by conjugation but, in contrast to their hosts, they have little or no capacity for oxidative detoxification (Munir and Barrett 1985). This is because helminths lack microsomal cytochrome systems (cytochromes P-450 and b_5). Therefore, the potential for developing resistance by detoxification is more limited in parasites than in their hosts.

This mechanism of resistance has not been investigated to any great extent in parasites. In a cambendazole-resistant strain of *H. contortus* the activity of a potential drug-metabolising enzyme, glutathione transferase was slightly higher than in the susceptible strain originally derived from the same population (Kawalek, Rew and Heavner 1984). But it is not clear whether this contributes to the resistance of the strain because we do not know whether this enzyme plays any role in detoxifying cambendazole in the parasite.

(h) *A change in development or behaviour within the host*
Drug pressure may select parasites that are genetically predisposed to inhabit sites where they happen to be less exposed to the compound, or to have a slightly different development cycle so that they tend to avoid exposure. There is evidence for *P. berghei* and *P. falciparum,* that parasites inhabiting reticulocytes are less susceptible to chloroquine than those in mature erythrocytes (Dei-Cas, Slomianny, Prensier, Vernes, Colin, Verhaeghe, Savage and Charet 1984). Therefore, chloroquine treatment could select for parasites preferentially infecting reticulocytes.

6.6   Conclusions

Biochemical variation is widespread in parasites and has been detected or is suspected in many species other than those illustrated above. It is almost impossible, without further work, to distinguish in many of these cases whether the variation is strictly genetic, or environmental, or a combination of the two. However, there is no doubt that a portion of the variation is genetic and that we are witnessing evolutionary processes in our laboratories even if this has little direct parallel with what happens in the field. It should be possible to exploit this situation to examine the host-parasite relationship and gain some insight into host influences on parasite biochemistry. It is essential, however, that genetic studies be done as well so that the contribution of selective and stochastic processes can be assessed and perhaps controlled for.

The implications of biochemical variation for the relative fitness of different strains or isolates may be important. At this stage, however, they are difficult to assess until such parameters as infectivity, growth rate, egg production, pathogenicity, etc., have been determined for strains with proven differences in rates of ATP synthesis or substrate utilisation. It is difficult to believe that the severe metabolic lesions found in the metronidazole-resistant laboratory strains of *T. foetus*, for example, have not rendered these strains less 'fit'. Their ability to compete with wild strains is probably reduced except when incubated *in vitro* in the presence of high concentrations of metronidazole. Thus 'fitness' is a matter of definition. 'Close adaptation' would be a more appropriate description for this case; flexibility has probably been lost and, as a consequence, 'overall' fitness, in a macroevolutionary sense, has probably also been lost.

Whether minor differences in apparent rates of ATP synthesis, as seen in most of the examples quoted above, reflect adaptive events – thus suggesting some metabolic compromise – or are the result of stochastic processes is an interesting question that cannot be answered at present. Some of the changes appear to be energetically neutral. Succinate or acetate production, for example, generate the same amount of ATP from glucose, so that, other things being equal, minor changes in the balance of these pathways would not affect the total yield of ATP. But a change from succinate or acetate to increased lactate production would either cost the parasite some ATP, which might be reflected in growth rate, for example, or would require the catabolism of additional quantities of substrate to make up the difference.

It should also be noted that parasites that have some functional classical aerobic respiration could be more affected by slight changes in the functioning of their aerobic pathways than by changes in the balance of their anaerobic pathways. This is because the yield of ATP is considerably higher from the aerobic pathways. Therefore, slight changes in the anaerobic pathways might, energetically speaking, be insignificant and perhaps not be subject to intense selection pressure. This depends, of course, whether the anaerobic pathways are located in critical tissues in the parasite in question, since not all tissues have access to sufficient oxygen to maintain aerobic respiration. However, since many parasites do have low levels of both TCA cycle activity and the classical electron transport pathways, these considerations should not be overlooked.

In this chapter we have been concerned with demonstrating that variation within species is the usual condition for living organisms including parasites. This conclusion is not new. Anyone with an understanding of evolutionary biology will ask why we have seen fit to document a biological truism through so many pages of text. The reason is that the phenomenon is a particularly important one in parasitology. Two features of the biology of parasites render them particularly susceptible to the evolutionary processes that induce variation. First, they are opportunistic, which is evident in the fact that most parasites have a great capacity for reproduction by sexual and asexual means. Second, they are exposed to the extreme selection procedures of the host-parasite relationship.

A knowledge of strain variation is important in studies of epidemiology and is essential for control measures.   It enables the rationalisation of conflicting physiological and biochemical data. Biochemical variation is made manifest by studies of genetics, nucleic acid analysis, enzyme polymorphism and metabolism, and we hope that future work will illuminate the ways in which these factors contribute to parasite fitness. Meanwhile, the literature on strain variation will continue in confusion **unless parasitologists specify, in much greater detail than is now the case, the host-parasite systems on which they work.**

Bibliography

Affranchino, J.L., De Tarlovsky, M.N.S. and Stoppani, A.O.M. (1985) 'Respiratory control in mitochondria from *Trypanosoma cruzi*' *Molecular and Biochemical Parasitology* **16**, 289-298

Affranchino, J.L., Schwarcz de Tarlovsky, M.N. and Stoppani, A.O.M. (1986) 'Terminal oxidases in the trypanosomatid *Trypanosoma cruzi*' *Comparative Biochemistry and Physiology* **85B**, 381-388

Aikawa, M. (1983) 'Host-parasite interaction: electron-microscopic study' in J. Guardiola, L. Luzzatto and W. Trager (eds), *Molecular Biology of Parasites* (Raven Press), pp 1-31

Aikawa, M., Miller, L.H., Johnson, J. and Rabbege, J.R. (1978) 'Erythrocyte entry by malarial parasites: a moving junction between erythrocyte and parasite' *Journal of Cell Biology* **77**, 72-82

Albert, A. (1985) *Selective Toxicity* 7th Ed. (Chapman and Hall, London)

Aldiss, B. (1982) *Helliconia Spring* (Jonathan Cape, London)

Almond, N.M. and Parkhouse, R.M.E. (1985) 'Nematode antigens' *Current Topics in Microbiology and Immunology* **120**, 173-203

Aman, R.A. and Wang, C.C. (1986) 'Absence of substrate chanelling in the glycosome of *Trypanosoma brucei*' *Molecular and Biochemical Parasitology* **19**, 1-10

Anders, R.F. (1986) 'Candidate antigens for an asexual blood-stage vaccine' *Parasitology Today* **1**, 152-155

Anderson, R.M. (1978) 'The regulation of host population growth by parasitic species' *Parasitology* **76**, 119-157

Anderson, R.M. and May, R.M. (1978) 'Regulation and stability of host-parasite population interactions. I. Regulatory processes' *Journal of Animal Ecology* **47**, 219-247

Anderson, R.M. and May, R.M. (1982) 'Coevolution of hosts and parasites' *Parasitology* **85**, 411-426

Anderson, R.M. and May, R.M. (1985a) 'Herd immunity to helminth infection: implications for parasite control' *Nature, London* **315**, 493-496

Anderson, R.M. and May, R.M. (1985b) 'Helminth infections of humans. Mathematical models, population dynamics and control' *Advances in Parasitology* **24**, 1-101

Andrews, P. and Thomas, H. (1979) 'Effect of praziquantel on *Hymenolepis diminuta in vitro*' *Tropenmedizin und Parasitologie* **30**, 391-400

Andrews, P., Thomas, H., Pohlke, R. and Seubert, J. (1983) 'Praziquantel' *Medicinal Research Reviews* **3**, 147-200

Arrick, B.A., Griffith, O.W. and Cerami, A. (1981) 'Inhibition of glutathione synthesis as a chemotherapeutic strategy for trypanosomiasis' *Journal of Experimental Medicine* **153**, 720-725

Askonas, B.A. (1984) 'Interference in general immune functions by parasite infections: African trypanosomiasis as a model system' *Parasitology* **88**, 633-638

Atkinson, D.E. (1971) 'Adenine nucleotides as stoichiometric coupling agents in metabolism and as regulatory modifiers: the adenylate energy charge' in H. Vogel (ed.), *Metabolic Pathways* Vol.5 (Academic Press, New York) pp 1-21

Aubry, M.L., Cowell, P., Davey, M.J. and Shevde, S. (1980) 'Aspects of the pharmacology of a new anthelmintic, pyrantel' *British Journal of Pharmacology* **38**, 332-344

Aurriault, C., Dessaint, J.P., Mazingue, C., Loyens, A. and Capron, A. (1984) 'Non-specific potentiation of T- and B-lymphocyte proliferation at the early stage of infection by *Schistosoma mansoni:* role of factors secreted by the larvae' *Parasite Immunology* **6**, 119-129

Avila, J.L., Avila, A. and Monzon, H. (1984) 'Differences in allopurinol and 4-aminopyrazolo(3,4-d)pyrimidine metabolism in drug-sensitive and insensitive strains of *Trypanosoma cruzi*' *Molecular and Biochemical Parasitology* **11**, 51-60

Avron, B., Deutsch, R.M. and Mirelman, D. (1982) 'Chitin synthesis inhibitors prevent cyst formation by *Entamoeba* trophozoites' *Biochemical and Biophysical Research Communications* **108**, 815-821

Ayala, F.J. (1983) 'Enzymes as taxonomic characters' in G.S. Oxford and D. Rollinson (eds) *Protein Polymorphism: Adaptive and Taxonomic Significance.* The Systematics Association Special Volume No.24 (Academic Press, London) pp 3-26

Bacchi, C.J. (1981) 'Content, synthesis and function of polyamines in trypanosomatids: relationship to chemotherapy' *Journal of Protozoology* **28**, 20-27

Badwey, J.A. and Karnovsky, M.L. (1980) 'Active oxygen species and the functions of phagocytic leukocytes' *Annual Review of Biochemistry* **49**, 695-726

Balloul, J.M., Sondermeyer, P., Dreyer, D., Capron, M., Grzych, J.M., Pierce, R.J., Carvallo, D., Lecocq, J.P. and Capron, A. (1987) 'Molecular cloning of a protective antigen of schistosomes' *Nature,London* **326**, 149-153

Barrett, J. (1981) *Biochemistry of Parasitic Helminths* (MacMillan, London)

Barrett, J. (1983) 'Biochemistry of filarial worms' *Helminthological Abstracts* **A52**, 1-18

Barrett, J. (1989) 'Parasitic helminths' in C. Bryant (ed.), *Metazoan Life without Oxygen.* (Chapman and Hall, London) (in press)

Barriga, O.O. (1981) *The Immunology of Parasitic Infections* (University Park Press, Baltimore).

Baverstock, P.R., Adams, M. and Beveridge, I. (1985) 'Biochemical differentiation in bile-duct cestodes and their marsupial hosts' *Molecular Biology and Evolution* **2**, 321-337

Beardsell, P.L. and Howell, M.J. (1984) 'Killing of *Taenia hydatigena* oncospheres by sheep neutrophils' *Zeitschrift für Parasitenkunde* **70**, 337-344

Bedi, A.J.K. and Isseroff, H. (1979) 'Bile duct hyperplasia in mice infected with *Schistosoma mansoni' International Journal for Parasitology* **9**, 401-404

Behm, C.A. and Bryant, C. (1976) 'Regulation of respiratory metabolism in *Moniezia expansa* under aerobic conditions' in H. Van den Bossche (ed.), *Biochemistry of Parasites and Host Parasite Relationships* (Elsevier/North Holland, Amsterdam) pp 89-94

Behm, C.A. and Bryant, C. (1985) 'The modes of action of some modern anthelmintics' in N. Anderson and P.J. Waller (eds), *Resistance in Nematodes to Anthelmintic Drugs* (CSIRO Animal Health/Australian Wool Corporation, Sydney) pp 57-67

Behm, C.A., Bryant, C. and Jones, A.J. (1987) 'Studies of glucose metabolism in *Hymenolepis diminuta* using ^{13}C nuclear magnetic resonance' *International Journal for Parasitology* **17**, 1333-1341

Bennet, E-M. and Bryant, C. (1984) 'Energy metabolism of adult *Haemonchus contortus* in vitro: a comparison of benzimidazole-susceptible and -resistant strains' *Molecular and Biochemical Parasitology* **10**, 335-346

Bennett, J.L. (1980) 'Characteristics of antischistosomal benzodiazepine binding sites in *Schistosoma mansoni*' *Journal of Parasitology* **66**, 742-747

Bennett, J.L. and Depenbusch, J.W. (1984) 'The chemotherapy of schistosomiasis' in J. Mansfield (ed.), *Parasitic Diseases* Vol. 2 (M. Dekker, New York) pp 73-131

Bernards, A., De Lange, T., Michels, P.A., Liu, A.Y.C., Huisman, M.J. and Borst, P. (1984) 'Two modes of activation of a single surface antigen gene of *Trypanosoma brucei*' *Cell* **36**, 163-170

Bhargava, K.K., Le Trang, N., Cerami, A. and Eaton, J.W. (1983) 'Effect of arsenical drugs on glutathione metabolism of *Litomosoides carinii*' *Molecular and Biochemical Parasitology* **9**, 29-35

Bienen, E.J., Hill, G.C. and Shin, K-O. (1983) 'Elaboration of mitochondrial function during *Trypanosoma brucei* differentiation' *Molecular and Biochemical Parasitology* **7**, 75-86

Blackwell, J., McMahon-Pratt, D. and Shaw, J. (1986) 'Molecular biology of *Leishmania*' *Parasitology Today* **2**, 45-53

Blair, R.J. and Weller, P.F. (1987) 'Uptake and esterification of arachidonic acid by trophozoites of *Giardia lamblia.*' *Molecular and Biochemical Parasitology* **25**, 11-18

Blum, J.J. (1987) 'Oxidation of fatty acids by *Leishmania braziliensis panamensis*' *Journal of Protozoology* **34**, 169-174

Blum, J.J., Yayon, A., Friedman, S. and Ginsburg, H. (1984) 'Effects of mitochondrial protein synthesis inhibitors on the incorporation of isoleucine into *Plasmodium falciparum in vitro*' *Journal of Protozoology* **31**, 475-479

Boddington, M.J. and Mettrick, D.F. (1981) 'Production and reproduction in *Hymenolepis diminuta* (Platyhelminthes: Cestoda)' *Canadian Journal of Zoology* **59**, 1962-1972

Bonay, P. and Cohen, B.E. (1983) 'Neutral amino acid transport in *Leishmania* promastigotes' *Biochimica et Biophysica Acta* **731**, 222-228

Bone, L.W. (1982) 'Chemotaxis of parasitic nematodes' in W.S. Bailey (ed.), *Cues that Influence Behavior of Internal Parasites* (USDA, New Orleans) pp 52-62

Boray, J.C. (1967) 'The life cycle of *Fasciola*' *Second International Liver Fluke Colloquium*, 11-16.

Borst, P. and Cross, G.A.M. (1982) 'Molecular basis for trypanosome antigenic variation' *Cell* **29**, 291-303

Borst, P., Fase-Fowler, F. and Gibson, W.C. (1981) 'Quantitation of genetic differences between *Trypanosoma brucei gambiense, rhodesiense* and *brucei* by restriction enzyme analysis of kinetoplast DNA' *Molecular and Biochemical Parasitology* **3**, 117-131

Boveris, A., Hertig, G. and Turrens, J.F. (1986) 'Fumarate reductase and other mitochondrial activities in *Trypanosoma cruzi.*' *Molecular and Biochemical Parasitology* **19**, 163-169

Bowlus, R.D. and Somero, G.N. (1979) 'Solute compatibility with enzyme function and structure; rationales for the selection of osmotic agents and end-products of anaerobic metabolism in marine invertebrates' *Journal of Experimental Zoology* **208**, 137-151

Bracha, R. and Mikelman, D. (1984) 'Virulence of *Entamoeba histolytica* trophozoites. Effects of bacteria, microaerobic conditions and metronidazole' *Journal of Experimental Medicine* **160**, 353-368

Branford White, C.J. and Hipkiss, J.B. (1984) 'The effect of calmodulin specific drugs on *Hymenolepis diminuta*' *Parasitology* **89**, lxxviii

Branford White, C.J., Hipkiss, J.B. and Peters, T.J. (1984) 'Evidence for a Ca^{2+}-dependent activator protein in the rat tapeworm *Hymenolepis diminuta.*' *Molecular and Biochemical Parasitology* **13**, 201-211

Brazier, J.B. and Jaffe, J.J. (1973) 'Two types of pyruvate kinase in schistosomes and filariae' *Comparative Biochemistry and Physiology* **44B**, 145-155

Brener, Z. (1973) 'Biology of *Trypanosoma cruzii*' *Annual Review of Microbiology* **27**, 347-382

Bricker, C.S., Depenbusch, J.W., Bennett, J.L. and Thompson, D.P. (1983) 'The relationship between tegumental disruption and muscle contraction in *Schistosoma mansoni* exposed to various compounds' *Zeitschrift für Parasitenkunde* **69**, 61-71

Broman, K., Knupfer, A-L., Ropars, M. and Deshusses, J. (1983) 'Occurrence and role of phosphoenolpyruvate carboxykinase in procyclic *Trypanosoma brucei brucei* glycosomes' *Molecular and Biochemical Parasitology* **8**, 79-87

Brown, K.N. (1969) 'Immunological aspects of malaria infection' *Advances in Immunology* **11**, 279-341

Brown, R.C., Evans, D.A. and Vickerman, K. (1973) 'Changes in oxidative metabolism and ultrastructure accompanying differentiation of the mitochondrion in *Trypanosoma brucei*' *International Journal for Parasitology* **3**, 691-704

Bryant, C. (1982) 'The biochemical origins of helminth parasitism' in L.E.A. Symons, A.D. Donald and J.K. Dineen (eds), *Biology and Control of Endoparasites* (Academic Press, Sydney) pp 29-52

Bryant, C. (1988) 'The biochemistry of *Dirofilaria immitis* ' in P.F.L.Boreham and R.B. Atwell (eds) *Dirofilariasis* (CRC Press, Boca Raton, Florida) pp 47-60

Bryant, C. (1989) 'Oxygen and the Lower Metazoa' in Bennet, E.M., Behm, C.A. and Bryant, C. (eds) *Comparative Biochemistry of Parasitic Helminths*, (Chapman and Hall, London) pp55-65

Bryant, C. and Bennet, E-M. (1983) 'Observations on the fumarate reductase system in *Haemonchus contortus* and their relevance to anthelmintic resistance and to strain variations of energy metabolism' *Molecular and Biochemical Parasitology* **7**, 281-292

Bryant, C. and Flockhart, H.A. (1986) 'Biochemical strain variation in parasitic helminths' *Advances in Parasitology* **25**, 276-319

Bueding, E. (1970) 'Biochemical effects of niridazole on *Schistosoma mansoni*' *Molecular Pharmacology* **6**, 532-539

Bueding, E. and Fisher, J. (1966) 'Factors affecting the inhibition of phosphofructokinase activity of *Schistosoma mansoni* by trivalent organic antimonials' *Biochemical Pharmacology* **15**, 1197-1211

Bueding, E., Dolan, P. and Leroy, J.P. (1982) 'The antischistosomal activity of oltipraz' *Research Communications in Chemical Pathology and Pharmacology* **37**, 293-303

Buteau, G.H., Simmons, T.E. and Fairbairn, D. (1969) 'Lipid metabolism in helminth parasites. IX. Fatty acid composition of shark tapeworms and of their hosts' *Experimental Parasitology* **26**, 209-213

Butterworth, A.E., Taylor, D.W., Veith, M.C., Vadas, M.A., Dessein, A., Sturrock, R.F. and Wells, E. (1982) 'Studies on the mechanisms of immunity in human schistosomiasis' *Immunological Reviews*, **61**, 5-39

Callow, L.L. and Dalgleish, R.J. (1982) 'Immunity and immunopathology in babesiosis' in S. Cohen and K.S. Warren (eds), *Immunology of Parasitic Infections* (Blackwell, Oxford), pp 475-526

Campbell, A.J. and Montague, P.E. (1981) 'A comparison of the activity of uncouplers of oxidative phosphorylation against the common liver fluke *Fasciola hepatica*' *Molecular and Biochemical Parasitology* **4**, 139-148

Campbell, W.C. (1985) 'Ivermectin: an update' *Parasitology Today* **1**, 10-16

Capron, A., Dessaint, J-P., Capron, M., Joseph, M. and Torpier, G. (1982) 'Effector mechanisms of immunity to schistosomes and their regulation' *Immunological Reviews* **61**, 41-66

Capron, M. and Capron, A. (1986) 'Rats, mice and men – models for immune effector mechanisms against schistosomiasis' *Parasitology Today* **2**, 69-75

Cerkasovová, A., Cerkasov, J. and Kulda, J. (1984) 'Metabolic differences between metronidazole resistant and susceptible strains of *Tritrichomonas foetus*' *Molecular and Biochemical Parasitology* **11**, 105-118

Chang, C.S. and Chang, K-P. (1985) 'Heme requirement and acquisition by extracellular and intracellular stages of *Leishmania mexicana amazonensis*' *Molecular and Biochemical Parasitology* **16**, 267-276

Chang, K.P. and Bray, R.S. (1985) *Leishmaniasis* (Elsevier, Amsterdam)

Cheah, K.S. (1972) 'Cytochromes in *Ascaris* and *Moniezia* ' in H. Van den Bossche (ed.), *Comparative Biochemistry of Parasites* (Academic Press, London) pp 417-432

Chensue, S.W., Kunkel, S.L., Higashi, G.I., Ward, P.A. and Boros, D.L. (1983) 'Production of superoxide anion, prostaglandins and hydroeicosatetraenoic acids from macrophages from hypersensitivity-type (*Schistosoma mansoni* egg) and foreign body-type granulomas' *Infection and Immunity* **42**, 1116-1125

Clark, I.A., Cowden, W.B. and Hunt, N.H. (1985) 'Free radical-induced pathology' *Medical Research Reviews* **5**, 297-332

Clark, I.A., Hunt, N.H. & Cowden, W.B. (1986) 'Oxygen derived free radicals in the pathogenesis of parasitic disease' *Advances in Parasitology* **25**, 1-45

Clarke, B. (1975) 'The contribution of ecological genetics to evolutionary theory: detecting the direct effects of natural selection on particular polymorphic loci' *Genetics* **79**, 101-113

Clegg, J.A. (1972) 'The schistosome surface in relation to parasitism' *Symposia of the British Society for Parasitology* **10**, 23-40

Clegg, J.A. (1974) 'Host antigens and the immune response in schistosomiasis' *CIBA Foundation Symposium* **25**, 161-183

Clegg, J.A., Smithers, S.R. & Terry, R.J. (1971) 'Acquisition of human antigens by *Schistosoma mansoni* during cultivation *in vitro*. *Nature, London* **232**, 653-654

Cohen, S. (1976) 'Survival of parasites in the immunized host' in S. Cohen and E. Sadun (eds), *Immunology of Parasitic Infections* (Blackwell, Oxford), pp 35-46

Cohen, S. (1979) 'Immunity to malaria' *Proceedings of the Royal Society of London, Series* **203**, 323-345

Cohen, S. (1984) 'Host-parasite interface: evasion' in K.S. Warren and A.A.F. Mahmoud (eds), *Tropical and Geographical Medicine* (McGraw-Hill), pp 138-146

Cohen, S. and Lambert, P.H. (1982) 'Malaria' in S. Cohen and K.S. Warren (eds), *Immunology of Parasitic Infections* (Blackwell, Oxford), pp 422-474

Cohen, S. and Sadun, E. (eds) (1976) *Immunology of Parasitic Infections.* (Blackwell, Oxford)

Cohen, S. and Warren, K.S. (eds) (1982) *Immunology of Parasitic Infections* (Blackwell, Oxford)

Coles, G.C. (1984) 'Recent advances in schistosome biochemistry' *Parasitology* **89**, 603-637

Coles, G.C, East, J.M. and Jenkins, S.N. (1975) 'The mechanism of action of the anthelmintic levamisole' *General Pharmacology* **6**, 309-313

Coles, G.C. and Simpkin, K.G. (1977) 'Metabolic gradient in *Hymenolepis diminuta* under aerobic conditions' *International Journal for Parasitology* **7**, 127-128

Colwell, R.K. (1973) 'Competition and coexistence in a simple tropical community' *American Naturalist* **107**, 737-760

Comley, J.C.W. (1985) 'Isoprenoid biosynthesis in filariae' *Tropical Medicine and Parasitology* **36** (Suppl.), 10-14

Comley, J.C.W. and Jaffe, J.J. (1983). 'The conversion of exogenous retinol and related compounds into retinyl phosphate mannose by adult *Brugia pahangi in vitro'*. *Biochemical Journal* **214**, 367-376

Comley, J.C.W., Jaffe, J.J. and Chrin, L.R. (1982) 'Glycosyl transferase activity in homogenates of *Dirofilaria immitis'* *Molecular and Biochemical Parasitology* **5**, 19-31

Coombs, G.C., Craft, J.A. and Hart, D.T. (1982) 'A comparative study of *Leishmania mexicana* amastigotes and promastigotes. Enzyme activities and subcellular locations' *Molecular and Biochemical Parasitology* **5**, 199-211

Coombs, G.H. (1986) 'Intermediary metabolism in parasitic protozoa' in M.J. Howell (ed.), *Parasitology – Quo Vadit?* Proceedings of the VI International Congress of Parasitology (Australian Academy of Science, Canberra) pp 97-104

Coombs, G.H. and Sanderson, B.E. (1985) 'Amine production by *Leishmania mexicana'* *Annals of Tropical Medicine and Parasitology* **79**, 409-416

Cornford, E.M. and Fitzpatrick, A.M. (1985) 'The mechanism and rate of glucose transfer from male to female schistosomes' *Molecular and Biochemical Parasitology* **17**, 131-141

Cowman, A.F., Timms, P. and Kemp, D.J. (1984) 'DNA polymorphisms and subpopulations in *Babesia bovis'* *Molecular and Biochemical Parasitology* **11**, 91-103

Cronin, C.N. and Tipton, K.F. (1985) 'Purification and regulatory properties of phosphofructokinase from *Trypanosoma (Trypanozoon) brucei brucei.'* *Biochemical Journal* **227**, 113-124

Curran, J., Baillie, D.L. and Webster, J.M. (1985) 'Use of genomic DNA restriction fragment length differences to identify nematode species' *Parasitology* **90**, 137-144

D'Alesandro, P.A. (1970) 'Non-pathogenic trypanosomes of rodents' in G.J. Jackson, R. Herman and I. Singer (eds), *Immunity to Parasitic Animals*, **2** (Appleton-Century-Crofts), pp 691-738

D'Alesandro, P.A. and Clarkson, A.B. (1980) '*Trypanosoma lewisi* : avidity and absorbability of ablastin, the rat antibody inhibiting parasite reproduction' *Experimental Parasitology* **50**, 384-396

Daddona, P.E., Wiesmann, W.P., Lambros, C., Kelley, W.N. and Webster, H.K. (1984) 'Human malaria parasite adenosine deaminase. Characterization in host enzyme-deficient erythrocyte culture' *Journal of Biological Chemistry* **259**, 1472-1475

Damien, R.T. (1964) 'Molecular mimicry: antigen sharing by parasite and host and its consequences' *American Naturalist* **98**, 129-143

Darling, T.N., Davis, D.G., London, R.E. and Blum, J.J. (1987) 'Products of *Leishmania braziliensis* glucose catabolism: release of D-lactate and, under anaerobic conditions, glycerol' *Proceedings of the National Academy of Sciences, USA* **84**, 7129-7133

Darwin, C. (1859) *The Origin of Species* (6th Edition, 1872) p. 643. John Murray, London

David, J. and Butterworth, A. (1979) 'Immunity to schistosomes. Advances and prospects' *Archives of Internal Medicine*, **91**, 641-643

David, J.R. (1984) 'Host-parasite interface: immunology' in K.S. Warren and A.A.F. Mahmoud (eds), *Tropical and Geographical Medicine* (McGraw-Hill, New York), pp 125-137

Davidse, L.C. and Flach, W. (1977) 'Differential binding of methyl benzimidazol-2-yl carbamate to fungal tubulin as a mechanism of resistance to this antimitotic agent in mutant strains of *Aspergillus nidulans*' *Journal of Cell Biology* **72**, 174-193

Dawson, P.J., Gutteridge, W.E. and Gull, K. (1983) 'Purification and characterisation of tubulin from the parasitic nematode, *Ascaridia galli*' *Molecular and Biochemical Parasitology* **7**, 267-277

Dei-Cas, E., Slomianny, C., Prensier, G. Vernes, A., Colin, J.J., Verhaeghe, A., Savage, A. and Charet, P. (1984) 'Action préférentielle de la chloroquine sur les *Plasmodium* hébergés dans des hématies matures' *Pathologie Biologie* **32**, 1019-1023

Desjardins, R.E. and Trenholme, G.M. (1984) 'Antimalarial chemotherapy' in J.M. Mansfield (ed.), *Parasitic Diseases*, Vol. 2 "The Chemotherapy" (M. Dekker, New York) pp 1-71

Despommier, D. (1976) 'Musculature' in C.R. Kennedy (ed.), ' *Ecological Aspects of Parasitology* ' (North Holland, Amsterdam) pp 269-285

Dineen, J.K. (1984) 'Immunological control of helminthiases by genetic manipulation of host and parasite' in J.K. Dineen and P.M. Outteridge (eds), *Immunogenetic Approaches to the Control of Endoparasites with Particular Reference to Parasites of Sheep* (CSIRO Publications), pp 1-9

Dineen, J.K. and Wagland, B.M. (1982) 'Immunoregulation of parasites in natural host-parasite systems – with special reference to the gastrointestinal nematodes of sheep' in L.E.A. Symons, A.D. Donald and J.K. Dineen (eds), *Biology and Control of Endoparasites* (Academic Press, Sydney), pp 297-329

Dineen, J.K., Gregg, P. and Lascelles, A.K. (1978) 'The response of lambs to vaccination at weaning with irradiated *Trichostrongylus colubriformis* larvae: segregation into 'responders' and 'non-responders' *International Journal for Parasitology* **8**, 59-63

Dinnick, J.A. and Dinnick, N.N. (1964) 'The influence of temperature on the succession of redial and cercarial generations of *Fasciola gigantica* in the snail host' *Parasitology* **54**, 59-65

Diribe, C.O. and Warhurst, D.C. (1985) 'A study of the uptake of chloroquine in malaria-infected erythrocytes. High and low affinity uptake and the influence of glucose and its analogues' *Biochemical Pharmacology* **34**, 3019-3027

Dixon, M. and Webb, E.C. (1979) *Enzymes* (Academic Press, New York)

Docampo, R. and Moreno, S.N.J. (1984) 'Free-radical intermediates in the antiparasitic action of drugs and phagocytic cells' in W.A. Pryor (ed.), *Free Radicals in Biology* Vol. VI (Academic Press, Orlando) pp243-288

Donahue, M.J., Masaracchia, R.A. and Harris, B.G. (1983) 'The role of cyclic AMP-mediated regulation of glycogen metabolism in levamisole-perfused *Ascaris suum* muscle' *Molecular Pharmacology* **23**, 378-383

Donelson, J.E. and Turner, M.J. (1985) 'How the trypanosome changes its coat' *Scientific American* **252**, 32-39

Edwards, S.R., Campbell, A.J., Sheers, M., Moore, R.J. and Montague, P.E. (1981) 'Studies of the effect of diamphenethide and oxyclozanide on the metabolism of *Fasciola hepatica.*' *Molecular and Biochemical Parasitology* **2**, 323-338

Ehrlich, P. (1909) 'Über den jetzigen Stand der Chemotherapie' *Chemische Berichte* **42**, 17-47

Eisenthal, R. and Panes, A. (1985) 'The aerobic/anaerobic transition of glucose metabolism in *Trypanosoma brucei' FEBS Letters* **181**, 23-27

El Kouni, M.H., Diop, D. and Cha, S. (1983) 'Combination therapy of schistosomiasis by tubercidin and nitrobenzylthioinosine 5'-monophosphate' *Proceedings of the National Academy of Sciences, USA* **80**, 6667-6670

El Kouni, M.H., Knopf, P.M. and Cha, S. (1985) 'Combination therapy of *Schistosoma japonicum* by tubercidin and nitrobenzylthioinosine 5'-monophosphate' *Biochemical Pharmacology* **34**, 3921-3924

Elford, B.C. (1986) 'L-glutamine influx in malaria-infected erythrocytes: a target for antimalarials?' *Parasitology Today* **2**(11), 309-312

Ellis, S.D., Li, Z.L, Gu, H.M., Peters, W., Robinson, B.L., Tovey, G. and Warhurst, D.C. (1985) 'The chemotherapy of rodent malaria, XXXIX. Ultrastructural changes following treatment with artemisinine of *Plasmodium berghei* infection in mice, with observations of the localization of [^{3}H]-dihydroartemisinine in *P. falciparum in vitro' Tropical Medicine and Parasitology* **79**, 367-374

Entner, N. (1979) 'Emetine binding to ribosomes of *Entamoeba histolytica* – inhibition of protein synthesis and amebicidal action' *Journal of Protozoology* **26**, 324-328

Ewald, P.W. (1983) 'Host-parasite relations, vectors and the evolution of disease severity' *Annual Review of Ecology and Systematics* **14**, 465-485

Fahey, R.C., Newton, G.L., Arrick, B., Overdank-Bogart, T. and Aley, S.B. (1984) '*Entamoeba histolytica* : a eukaryote without glutathione metabolism' *Science* **224**, 70-72

Fairbairn, D. (1970) 'Biochemical adaptation and loss of genetic capacity in helminth parasites' *Biological Reviews* **45**, 29-72

Fairbairn, D., Wertheim, G., Harpur, R.P. and Schiller, E.L. (1961) 'Biochemistry of normal and irradiated strains of *Hymenolepis diminuta' Experimental Parasitology* **11**, 248-263

Fairlamb, A.H. (1981) 'Discussion' in R.A. Klein and P.G.G. Miller (eds), 'Alternate Metabolic Pathways in Protozoan Energy Metabolism' in 'Workshop Proceedings of the 3rd European Multicolloquium of Parasitology', *Parasitology* **82**, 22-23

Fairlamb, A.H. (1982) 'Biochemistry of trypanosomiasis and rational approaches to chemotherapy' *Trends in Biochemical Sciences* **7**, 249-253

Fairlamb, A.H. and Bowman, I.B.R. (1980) 'Uptake of the trypanocidal drug suramin by bloodstream forms of *Trypanosoma brucei* and its effect on respiration and growth rate *in vivo' Molecular and Biochemical Parasitology* **1**, 315-333

Fairlamb, A.H., Blackburn, P., Ulrich, P., Chait, B.T. and Cerami, A. (1985) 'Trypanothione: a novel bis(glutathionyl) spermidine cofactor for glutathione reductase in trypanosomatids' *Science* **227**, 1485-1487

Fairweather, I., Holmes, S.D. and Threadgold, L.T. (1984) '*Fasciola hepatica* : motility response to fasciolicides *in vitro' Experimental Parasitology* **57**, 209-224

Fantone, J.C. and Ward, P.A. (1982) 'Role of oxygen-derived free radical metabolites in leukocyte-dependent inflammatory reactions' *American Journal of Pathology* **107**, 397-417

Fitch, C.D., Dutta, P., Kanjananggulpan, P. and Chevli, R. (1984) 'Ferriprotoporphyrin IX: a mediator of the antimalarial action of oxidants and 4-aminoquinoline drugs' in J.W. Eaton and G.J. Brewer (eds), *Malaria and the Red Cell*, Progress in Clinical and Biological Research, Vol. 155, (Alan R. Liss, New York) pp 119-130

Flockhart, H.A., Cibulskis, R., Karam, M. and Albiez, E.J. (1986) 'Onchocerca volvulus : enzyme polymorphism in relation to the differentiation of forest and savannah strains of the parasite' Transactions of the Royal Society of Tropical Medicine and Hygiene 80, 285-292

Flockhart, H.A., Harrison, S.E., Dobinson, A.R. and James, E.R. (1982) 'Enzyme polymorphism in Trichinella.' Transactions of the Royal Society of Tropical Medicine and Hygiene 76, 541-545

Frank, W. (1982) (ed.), Immune Reactions to Parasites (Gustav Fischer Verlag, Stüttgart)

Frayha, Q.J. and Smyth, J.D. (1983) 'Lipid metabolism in parasitic helminths' Advances in Parasitology 22, 309-387

Freeland, W.J. (1983) 'Parasites and the coexistence of animal host species' American Naturalist 121, 223-236

Friedman, M.J. and Trager, W. (1981) 'The biochemistry of resistance to malaria' Scientific American 244, 154-164.

Fry, M. and Brazeley, E.P. (1985) 'Mitochondrial NADH-fumarate reductase in adult Brugia pahangi (Abstract) Tropenmedizin und Parasitologie 36, (Suppl.) p. 25

Fry, M. and Jenkins, D.C. (1984) 'Nematoda: aerobic respiratory pathways of adult parasitic species' Experimental Parasitology 57, 86-92

Fuhrman, J.A. and Piessens, W.F. (1985) 'Chitin synthesis and sheath morphogenesis in Brugia malayi microfilariae' Molecular and Biochemical Parasitology 17, 93-104

Game, S., Holman, G. and Eisenthal, R. (1986) 'Sugar transport in Trypanosoma brucei a suitable kinetic probe.' FEBS Letters 194, 126-130

Geary, T.G., Delaney, E.J., Klotz, I.M. and Jensen, J.B. (1983) 'Inhibition of the growth of Plasmodium falciparum in vitro by covalent modification of hemoglobin' Molecular and Biochemical Parasitology 9, 59-72

Geczy, A.F. (1984) 'Genetic susceptibility to disease in humans with particular reference to some parasitic infections' in J.K. Dineen and P.M. Outteridge (eds), Immunogenetic Approaches to the Control of Endoparasites with Particular Reference to Parasites of Sheep (CSIRO Publications), pp 93-102

Gibson, W., Borst, P. and Fase-Fowler, F. (1985) 'Further analysis of intraspecific variation in Trypanosoma brucei using restriction site polymorphisms in the maxicircle of kinetoplast DNA' Molecular and Biochemical Parasitology 15, 21-36

Gibson, W.C., Marshall, T.F.deC. and Godfrey, D.G. (1980) 'Numerical analysis of enzyme polymorphisms: new approach to the epidemiology and taxonomy of trypanosomes of the subgenus Trypanozoon' Advances in Parasitology 18, 175-246

Ginsburg, H., Divo, A.A., Geary, T.G., Boland, M.T. and Jensen, J.B.(1986) 'Effects of mitochondrial inhibitors on intraerythrocytic Plasmodium falciparum in in vitro cultures' Journal of Protozoology 33, 121-125

Ginsburg, H., Kutner, S., Krugliak, M. and Cabantchik, Z.I. (1985) 'Characterization of permeation pathways appearing in the host membrane of Plasmodium falciparum infected red blood cells' Molecular and Biochemical Parasitology 14, 313-322

Goad, L.J., Holz, G.G. and Beach, D.H. (1985) 'Sterols of ketoconazole-inhibited Leishmania mexicana mexicana promastigotes' Molecular and Biochemical Parasitology 15, 257-279

Goijman, S.G., Frasch, A.C.C. and Stoppani, A.O.M. (1985) 'Damage of Trypanosoma cruzi deoxyribonucleic acid by nitroheterocyclic drugs' Biochemical Pharmacology 34, 1457-1461

Goldberg, E. (1966) 'Lactate dehydrogenase of trout: hybridisation *in vivo* and *in vitro*' *Science* **151**, 1091-1093

Gould, S.J. and Lewontin, R.C. (1979) 'The spandrels of San Marco and the Panglossian paradigm: a critique of the adaptationist program' *Proceedings of the Royal Society of London, Series B* **205**, 581-598

Giffin, B.F., McCann, P.P., Bitonti, A.J. and Bacchi, C.J. (1986) 'Polyamine depletion following exposure to DL-alpha-difluoromethylornithine both *in vivo* and *in vitro* initiates morphological alterations and mitochondrial activation in a monomorphic strain of *Trypanosoma brucei brucei*' *Journal of Protozoology* **33**, 238-242

Grillo, M.A. (1985) 'Metabolism and function of polyamines' *International Journal of Biochemistry* **17**, 943-948

Gugliotta, J.L., Tanowitz, H.B., Wittner, M. and Soeiro, R. (1980) '*Trypanosoma cruzi*: inhibition of protein synthesis by nitrofuran SQ 18,506' *Experimental Parasitology* **49**, 216-224

Gutteridge, W.E., Dave, D. and Richards, W.H.G. (1979) 'Conversion of dihyroorotate to orotate in parasitic protozoa' *Biochimica et Biophysica Acta* **582**, 390-401

Hajduk, S.L. (1984) 'Antigenic variation during the developmental cycle of *Trypanosoma brucei*' *Journal of Protozoology*, **31**, 41-47

Halliwell, B. and Gutteridge, J.M.C. (1985) *Free Radicals in Biology and Medicine* (Clarendon Press, Oxford)

Hamilton, W.D. (1980) 'Sex versus non-sex versus parasite' *Oikos* **35**, 282-290

Hamilton, W.D. (1982) 'Pathogens as causes of genetic diversity in their host populations' in R.M. Anderson and R.M. May (eds), *Population Biology of Infectious Diseases* (Springer-Verlag), pp 269-296

Hammond, D.J. and Gutteridge, W.E. (1984) 'Purine and pyrimidine metabolism in the trypanosomatidae' *Molecular and Biochemical Parasitology* **13**, 243-261

Hammond, D.J., Burchell, J.R. and Pudney, M. (1985) 'Inhibition of pyrimidine biosynthesis *de novo* in *Plasmodium falciparum* by 2-(4-t-butylcyclohexyl)-3-hydroxy-1,4-naphthoquinone *in vitro*' *Molecular and Biochemical Parasitology* **14**, 97-109

Hart, D.T., Misset, O., Edwards, S.W. and Opperdoes, F.R. (1984) 'A comparison of the glycosomes (microbodies) isolated from *Trypanosoma brucei* bloodstream form and cultured procyclic trypomastigotes' *Molecular and Biochemical Parasitology* **12**, 25-35

Hart, D.T., Vickerman, K. and Coombs, G.C. (1981) 'Respiration of *Leishmania mexicana* amastigotes and promastigotes' *Molecular and Biochemical Parasitology* **4**, 39-51

Henderson, J.F., Zombor, G., Johnson, M.M. and Smith, C.M. (1983) 'Variation in erythrocyte purine metabolism among mouse strains' *Comparative Biochemistry and Physiology* **76B**, 419-422

Hennessy, D.R. (1985) 'Manipulation of anthelmintic pharmacokinetics' in N. Anderson and P.J. Waller (eds) *Resistance in Nematodes to Anthelmintic Drugs* (CSIRO Animal Health/Australian Wool Corporation, Sydney) pp 79-85

Heyworth, P.G., Gutteridge, W.E. and Ginger, C.D. (1984) 'Pyrimidine metabolism in *Trichomonas vaginalis*' *FEBS Letters* **176**, 55-60

Hochachka, P.W. and Somero, G.N. (1984) *Biochemical Adaptation* (Princeton University Press, Princeton)

Hofer, H.W., Allen, B.J., Kaeini, M.R. and Harris, B.G. (1982) 'Phosphofructokinase from *Ascaris suum*. The effect of phosphorylation on activity at near-physiological conditions' *Journal of Biological Chemistry* **257**, 3807-3810

Hollingdale, M.R., McCann, P.P. and Sjoerdsma, A. (1985) '*Plasmodium berghei* : inhibitors of ornithine decarboxylase block exoerythrocyte schizogony' *Experimental Parasitology* 60, 111-117

Holmes, J.C. and Price, P.W. (1986) 'Communities of parasites' in J. Kikkawa and D.J. Anderson (eds), '*Community Ecology: Pattern and Process.*' (Blackwell, Oxford) pp 187-213

Howard, R.J. (1986) 'Vaccination against malaria: recent advances and the problems of antigenic diversity and other parasite evasion mechanisms' in M.J. Howell (ed.), *Parasitology - Quo Vadit?* (Australian Academy of Science), pp 17-29

Howard, R.J. and Miller, L.H. (1981) 'Invasion of erythrocytes by malaria merozoites: evidence for specific receptors involved in attachment and entry' in *CIBA Foundation Symposium* 80, "Adhesion and Microorganism Pathogenicity" (Pittman Medical), pp 202-219

Howard, R.J., Panton, L.J., Marsh, K., Winchell, E.J. and Wilson, R.J.M. (1986) 'Antigenic diversity and size diversity of *Plasmodium falciparum* antigens in isolates from Gambian patients. I. S-antigens.' *Parasite Immunology* 8, 39-55

Howell, M.J. (1976) 'The peritoneal cavity of vertebrates' in C.R. Kennedy (ed.), *Ecological Aspects of Parasitology* (North-Holland, Amsterdam) pp 243-268

Howell, M.J. (1985) 'Gene exchange between hosts and parasites' *International Journal for Parasitology* 15, 597-600

Howell, M.J. (1986) 'Cultivation of *Echinococcus* species *in vitro*' in R.C.A. Thompson (ed.), *The Biology of Echinococcus and Hydatid Disease* (George Allen and Unwin, London) pp 143-163

Howells, R.E. (1985) 'The modes of action of some anti-protozoal drugs' *Parasitology* 90, 687-704

Howells, R.E., Mendis, A.M. and Bray, P.G. (1983) 'The mode of action of suramin on the filarial worm *Brugia pahangi*' *Parasitology* 87, 29-48

Howells, R.E., Tinsley, J., Devaney, E. and Smith, G. (1981) 'The effect of 5-fluorouracil and 5-fluorocytosine on the development of the filarial nematodes *Brugia pahangi* and *Dirofilaria immitis*' *Acta Tropica* 38, 289-304

Hsu, W.H. (1980) 'Toxicity and drug interactions of levamisole' *Journal of the American Veterinary Medical Association* 176, 1166-1169

Huang, T-Y. (1980) 'The energy metabolism of *Schistosoma japonicum.*' *International Journal of Biochemistry* 12, 457-464

Hudson, A.T., Randall, A.W., Fry, M., Ginger, C.D., Hill, B., Latter, V.S., McHardy, N. and Williams, R.B. (1985) 'Novel anti-malarial hydroxynaphthoquinones with potent broad spectrum anti-protozoal activity' *Parasitology* 90, 45-55

Hudson, L. (ed.) (1985) , *The Biology of Trypanosomes* (Springer-Verlag, Berlin)

Hurd, H. and Arme, C. (1984) 'Pathophysiology of *Hymenolepis diminuta* infections in *Tenebrio molitor*: effect of parasitism on haemolymph proteins' *Parasitology* 89, 253-262

Hyman, L.H. (1919) 'On the action of certain substances on oxygen consumption' *American Journal of Physiology* 48, 340-371

Iltzsch, M.H., Niedzwicki, J.G., Senft, A.W., Cha, S. and El Kouni, M.H. (1984) 'Enzymes of uridine 5'-monophosphate biosynthesis in *Schistosoma mansoni*' *Molecular and Biochemical Parasitology* 12, 153-171

Inselberg, J. (1983) 'Stage-specific inhibitory effect of cyclic AMP on asexual maturation and gametocyte formation of *Plasmodium falciparum*' *Journal of Parasitology* 69, 592-597

Iovannisci, D.M. and Ullman, B. (1984) 'Characterization of a mutant *Leishmania donovani* deficient in adenosine kinase activity' *Molecular and Biochemical Parasitology* 12, 139-151

Iovannisci, D.M., Goebel, D., Allen, K., Kaur, K. and Ullman, B. (1984) 'Genetic analysis of adenine metabolism in *Leishmania donovani* promastigotes. Evidence for diploidy at the adenine phosphoribosyltransferase locus' *Journal of Biological Chemistry* **259**, 14617-14623

Isseroff, H., Girard, P.R. and Leve, M.D. (1977) 'Bile duct enlargement induced in rats after intraperitoneal transplantation' *Experimental Parasitology* **41**, 405-409

Isseroff, H., Sawma, J.T. and Raino D. (1977) 'Fascioliasis: role of proline in bile duct hyperplasia' *Science* **198**, 1157-1159

Jack, R.M. and Ward, P.A. (1981) 'Mechanism of entry of *Plasmodia* and *Babesia* into red cells' in M. Ristic and J.P. Kreier (eds), *Babesiosis* (Academic Press, New York) pp 445-457

Jaffe, J.J. (1972) 'Dihydrofolate reductases in parasitic protozoa and helminths' in H. Van den Bossche (ed.), *Comparative Biochemistry of Parasites* (Academic Press, New York) pp 219-233

Jaffe, J.J. (1980) 'Filarial folate-related metabolism is a potential target for selective inhibitors' in H. Van den Bossche (ed.), *The Host-Invader Interplay* (Elsevier/North Holland, Amsterdam) pp 605-614

Jaffe, J.J. (1981) 'Involvement of tetrahydrofolate cofactors in *de novo* purine ribonucleotide synthesis by adult *Brugia pahangi* and *Dirofilaria immitis*' *Molecular and Biochemical Parasitology* **2**, 259-270

Jaffe, J.J. and Chrin, L.R. (1980) 'Folate metabolism in filariae: enzymes associated with 5,10-methylenetetrahydrofolate' *Journal of Parasitology* **66**, 53-58

Jaffe, J.J., Chrin, L.R. and Smith, R.B. (1980) 'Folate metabolism in filariae. Enzymes associated with 5,10-methenyltetrahydrofolate and 10-formyltetrahydrofolate' *Journal of Parasitology* **66**, 428-433

James, S. (1980) 'Thiamine uptake in isolated schizonts of *Eimeria tenella* and the inhibitory effects of amprolium' *Parasitology* **80**, 313-322

Jansma, W.B., Rogers, S.H., Lin, C.L. and Bueding, E. (1977) 'Experimentally produced resistance of *Schistosoma mansoni* to hycanthone' *American Journal of Tropical Medicine and Hygiene* **26**, 926-936

Jarroll, E.L., Hammond, M.M. and Lindmark, D.G. (1987) '*Giardia lamblia*. Uptake of pyrimidine nucleosides' *Experimental Parasitology* **63**, 152-156

Johnson, R. (1982) 'Parsimony principles in phylogenetic systematics: a critical appraisal' *Evolutionary Theory* **6**, 79-90

Kan, S.C. and Siddiqui, W.A. (1979) 'Comparative studies on dihydrofolate reductases from *Plasmodium falciparum* and *Aotus trivirgatus*' *Journal of Protozoology* **26**, 660-664

Karvonen, E., Kauppinen, L., Partanen, T. and Pösö, H. (1985) 'Irreversible inhibition of putrescine-stimulated S-adenosyl-L-methionine decarboxylase by Berenil and Pentamidine' *Biochemical Journal* **231**, 165-169

Kass, I.S., Stretton, A.O.W. and Wang, C.C. (1984) 'The effects of avermectin and drugs related to acetylcholine and 4-aminobutyric acid on neurotransmission in *Ascaris suum*' *Molecular and Biochemical Parasitology* **13**, 213-225

Kassai, T. (1982) *Handbook of Nippostrongylus brasiliensis (Nematode)* Commonwealth Agricultural Bureau, London.

Kawalek, J.C., Rew, R.S. and Heavner, J. (1984) 'Glutathione-S-transferase, a possible drug-metabolizing enzyme, in *Haemonchus contortus*: comparative activity of a cambendazole-resistant and a susceptible strain' *International Journal for Parasitology* **14**, 173-175

Kelley, W.N. (1983) 'Hereditary orotic aciduria'. in J.B. Stanbury, J.B. Wyngaarden, D.S. Frederiksen, J.L. Goldstein and M.S. Brown (eds), *The Metabolic Basis of Inherited Disease* 5th Ed., McGraw-Hill, New York, pp 1202-1226

Kemp, D.J., Coppel, R.L., Stahl, H.D., Bianco, A.E., Corcoran, L.M., McIntyre, P., Langford, C.J., Favaloro, J.M., Crewther, P.E., Brown, G.V., Mitchell, G.F., Culvenor, J.G. and Anders, R.F. (1986) 'Genes for antigens of *Plasmodium falciparum*' *Parasitology* **91**, S83-S108

Kidder, G.W. and Nolan, L.L. (1981) 'The *in vivo* and *in vitro* action of 4-amino-5-imidazolecarboxamide in trypanosomatid flagellates' *Molecular and Biochemical Parasitology* **3**, 265-270

Kimura, M. (1983) *The Neutral Theory of Molecular Evolution* (Cambridge University Press)

Kita, K., Takamiya, S., Furushima, R., Ma, Y-C. and Oya, H. (1988) Complex II is a major component of the respiratory chain in the muscle mitochondria of *Ascaris suum* with high fumarate reductase activity' *Comparative Biochemistry and Physiology* **89B**, 31-34

Kitchener, K.R., Meshnick, S.R., Fairfield, A.S. and Wang, C.C. (1984) 'An iron-containing superoxide dismutase in *Tritrichomonas foetus*' *Molecular and Biochemical Parasitology* **12**, 95-99

Klein, J. (1982) *Immunology – The Science of Self-Non-Self Discrimination* (John Wiley, New York)

Koehne, R.K. (1978) 'Physiology and biochemistry of enzyme variation: the interface of ecology and population genetics' in P.D. Bussard (ed.) *Ecological Genetics: The Interface* (Springer, Berlin) pp 51-72

Köhler, P. (1985) 'The strategies of energy conservation in helminths' *Molecular and Biochemical Parasitology* **17**, 1-18

Köhler, P. and Bachmann, R. (1978) 'The effects of the antiparasitic drugs levamisole, thiabendazole, praziquantel, and chloroquine on mitochondrial electron transport in muscle tissue from *Ascaris suum*' *Molecular Pharmacology* **14**, 155-163

Kohlhagen, S. (1988) PhD Thesis, Australian National University

Kohlhagen, S. and Bryant, C. 'Aspects of the metabolism of *Taenia serialis* ' (in preparation)

Kohlhagen, S., Behm, C.A. and Bryant, C. (1985) 'Strain variation in *Hymenolepis diminuta*: enzyme profiles' *International Journal for Parasitology* **15**, 479-483

Komuniecki, P.R. and Saz, H.J. (1982) 'The effect of levamisole on glycogen synthase and the metabolism of *Litomosoides carinii*' *Journal of Parasitology* **68**, 221-227

Komuniecki, R., Komuniecki, P.R. and Saz, H.J. (1981a) 'Relationship between pyruvate decarboxylation and branched-chain volatile acid synthesis in *Ascaris* mitochondria' *Journal of Parasitology* **67**, 601-608

Komuniecki, R., Komuniecki, P.R. and Saz, H.J. (1981b) 'Pathway of formation of branched-chain volatile fatty acids in *Ascaris* mitochondria' *Journal of Parasitology* **67**, 841-846

Komuniecki, R., Wack, M. and Coulson, M. (1983) 'Regulation of the *Ascaris suum* pyruvate dehydrogenase complex by phosphorylation and dephosphorylation' *Molecular and Biochemical Parasitology* **8**, 165-176

Königk, E. and Putfarken, B. (1985) 'Ornithine decarboxylase of *Plasmodium falciparum*: a peak-function enzyme and its inhibition by chloroquine' *Tropical Medicine and Parasitology* **36**, 81-84

Konings, W.N. (1985) 'Generation of metabolic energy by end-product efflux' *Topics in Biochemical Science* **10**, 317-319

Koshland, D.E. (1984) 'Control of enzyme activity and metabolic pathways' *Trends in Biochemical Sciences* **9**, 155-159

Krahenbuhl, J.L. and Remington, J.S. (1982) 'The immunology of *Toxoplasma* and toxoplasmosis' in S. Cohen and K.S. Warren (eds), *Immunology of Parasitic Infections* (Blackwell, Oxford), pp 356-421

Krogsgaard-Larsen, P. (1981) 'Gamma-aminobutyric acid agonists, antagonists, and uptake inhibitors. Design and therapeutic aspects' *Journal of Medicinal Chemistry* **24**, 1377-1383

Kulda, J., Cerkasov, J. and Demes, P. (1984) '*Tritrichomonas foetus* : stable anaerobic resistance to metronidazole *in vitro*' *Experimental Parasitology* **57**, 93-103

Kurelec, B. (1975) 'Catabolic path of arginine and NAD regeneration in the parasite *Fasciola hepatica*' *Comparative Biochemistry and Physiology* **51B**, 151-156

Kuwahara, T., White, R.A. and Agosin, M. (1985) 'A cytosolic FAD-containing enzyme catalysing cytochrome c reduction in *Trypanosoma cruzi* . I. Purification and some properties' *Archives of Biochemistry and Biophysics* **239**, 18-28

Lacey, E. (1985) 'The biochemistry of anthelmintic resistance' in N. Anderson and P.J. Waller (eds), *Resistance in Nematodes to Anthelmintic Drugs* (CSIRO Animal Health/Australian Wool Corporation, Sydney) pp 69-78

Laurie, J.S. (1957) 'The *in vitro* fermentation of carbohydrates by two species of cestodes and one species of Acanthocephala' *Experimental Parasitology* **6**, 245-260

Lavie, B. and Nevo, E. (1982) 'Heavy-metal selection of phosphoglucose isomerase allozymes in marine gastropods' *Marine Biology* **71**, 17-22

Le Jambre, L.F. (1985) 'Genetic aspects of anthelmintic resistance in nematodes' in N. Anderson and P.J. Waller (eds), *Resistance in Nematodes to Anthelmintic Drugs* (CSIRO Animal Health/Australian Wool Corporation, Sydney) pp 97-106

Leitch,, B. and Probert, A.J. (1984) '*Schistosoma haematobium*: amoscanate and adult worm ultrastructure' *Experimental Parasitology* **58**, 278-289

Levin, B.R., Allison, A.C., Bremermann, H.J., Cavill-Sforza, Clarke, B.C., Frentzel-Beyme, R., Hamilton, W.D., Levin, S.A., May, R.M. and Thieme, H.R. (1982) 'Evolution of parasites and hosts' in R.M. Anderson and R.M. May (eds), *Population Biology of Infectious Diseases* (Springer-Verlag), pp 213-243

Lewis, J.A., Wu, C.H., Levine, J.H. and Berg, H. (1980) 'Levamisole-resistant mutants of the nematode *Caenorhabditis elegans* appear to lack pharmacological acetylcholine receptors' *Neuroscience* **5**, 967-989

Lindmark, D.G. (1980) 'Energy metabolism of the anaerobic protozoon *Giardia lamblia*' *Molecular and Biochemical Parasitology* **1**, 1-12

Lloyd, D. and Kristensen, B. (1985) 'Metronidazole inhibition of hydrogen production *in vivo* in drug-sensitive and resistant strains of *Trichomonas vaginalis*' *Journal of General Microbiology* **131**, 849-853

Lloyd, D. and Pedersen, J.Z. (1985) 'Metronidazole radical anion generation *in vivo* in *Trichomonas vaginalis*: oxygen quenching is enhanced in a drug-resistant strain' *Journal of General Microbiology* **131**, 87-92

Loo, V.G. and Lalonde, R.G. (1984) 'Role of iron in intracellular growth of *Trypanosoma cruzi*' *Infection and Immunity* **45**, 726-730

Lumsden, R.D. and Murphy, W.A. (1980) 'Morphological and functional aspects of the cestode surface' in C.B. Cook., P.W. Pappas and E.D. Rudolph (eds), *Cellular Interactions in Symbiosis and Parasitism* (Ohio State University Press, Columbus) pp 95-130

MacInnis, A.J. (1976) 'How parasites find hosts: some thoughts on the inception of the host-parasite integration' in C.R. Kennedy (ed.), *Ecological Aspects of Parasitology* (North Holland, Amsterdam)

Mackenzie, N.E., Hall, J.E., Flynn, I.W. and Scott, A.I. (1983) '[13]C nuclear magnetic resonance studies of anaerobic glycolysis in *Trypanosoma brucei* spp' *Bioscience Reports* **3**, 141-151

Mahoney, D.F. (1977) '*Babesia* of domestic animals' in J.P. Kreier (ed.), *Parasitic Protozoa* Vol. iv . (Academic Press), pp 1-52.

Maizels, R.M., Philipp, M. and Ogilvie, B.M. (1982) 'Molecules on the surface of parasitic nematodes as probes of the immune response to infection' *Immunological Reviews* 61, 109-136

Mansfield, J.M. (ed.) (1981), *Parasitic Diseases*. Vol.1 'The Immunology' (Marcel-Dekker,New York).

Mansour, T.E. (1984) 'Serotonin receptors in parasitic worms' *Advances in Parasitology* 23, 1-36

Margulies, L. (1981) '*Symbiosis in Cell Evolution*' (W.H. Freeman, San Francisco)

Marr, J.J. (1984) 'The chemotherapy of leishmaniasis' in J.M. Mansfield (ed.), *Parasitic Diseases*. Vol. 2, 'The Chemotherapy' (M. Dekker, New York) pp 201-227

Marrack, P. and Kappler, J. (1986) 'The T cell and its receptor' *Scientific American* 254, 28-37

Massamba, N.N. and Williams, R.O. (1984) 'Distinction of African trypanosome species using nucleic acid hybridisation' *Parasitology* 88, 55-65

Matthews, P.M., Foxall, D., Shen, L. and Mansour, T.E. (1986) 'Nuclear magnetic resonance studies of carbohydrate metabolism and substrate cycling in *Fasciola hepatica*' *Molecular Pharmacology* 29, 65-73

Mattoccia, L.P., Lelli, A. and Cioli, D. (1981) 'Effect of hycanthone on *Schistosoma mansoni* macromolecular synthesis *in vitro*' *Molecular and Biochemical Parasitology* 2, 295-308

Mauel, J. (1984) 'Mechanisms of survival of protozoan parasites in mononuclear phagocytes' *Parasitology* 88, 579-592

Mauel, J. and Behin, R. (1982) 'Leishmaniasis' in S. Cohen and K.S. Warren (eds), *Immunology of Parasitic Infections* (Blackwell, Oxford), pp 299-355

May, R.M. and Anderson, R.M. (1983) 'Epidemiology and genetics in the coevolution of parasites and hosts' *Proceedings of the Royal Society of London, Series B* 219, 281-313

Mayr, E. (1983) 'How to carry out the adaptationist program' *American Naturalist* 121, 324-334

McAlister, R.O. and Mishra, G.C. (1983) 'Putative inhibitors of erythrocyte transmembrane Ca^{2+} kill *Plasmodium falciparum in vitro*' *Journal of Parasitology* 69, 777-778

McCabe, R.E., Remington, J.S. and Araujo, F.G. (1984) 'Ketoconazole inhibition of intracellular multiplication of *Trypanosoma cruzi* and protection of mice against lethal infection with the organism' *Journal of Infectious Diseases* 150, 594-601

McCann, P.P., Bacchi, C.J., Nathan, H.C. and Sjoerdsma, A. (1983) 'Difluoromethylornithine and the rational development of polyamine antagonists for the cure of protozoan infection' in T.P. Singer and R.N. Ondarza *Mechanisms of Drug Action* (Academic Press, New York) pp 159-173

McCormack, J.J. (1981) 'Dihydrofolate reductase inhibitors as potential drugs' *Medicinal Research Reviews* 1, 303-331

McIntyre, P., Coppel, R.L., Smith, D.B., Stahl, H.D., Corcoran, L.M., Langford, C.J., Favaloro, J.M., Crewther, P.E., Brown, G.V., Mitchell, G.F., Anders, R.F. and Kemp, D.J. (1986) 'Expression of parasite antigens in *Escherichia coli* ' in M.J. Howell (ed.), *Parasitology – Quo Vadit?* (Australian Academy of Science), pp 59-67

McKenzie, J.A. (1985) 'The genetics of resistance to chemotherapeutic agents' in N. Anderson and P.J. Waller (eds), *Resistance in Nematodes to Anthelmintic Drugs* (CSIRO Animal Health/Australian Wool Corporation, Sydney) pp 89-95

McLaren, D.J. (1980) *Schistosoma mansoni: The Parasite Surface in Relation to Host Immunity* (John Wiley, New York)

McLaren, D.J. (1984) 'Disguise as an evasion strategem of parasitic organisms' *Parasitology* 88, 597-611

McLaughlin, J. and Aley, S. (1985) 'The biochemistry and functional morphology of the *Entamoeba*' *Journal of Protozoology* 32, 221-240

McManus, D.P. (1981) 'A biochemical study of adult and cystic stages of *Echinococcus granulosus* of human and animal origin from Kenya' *Journal of Helminthology* 55, 21-27

McManus, D.P. and Smyth, J.D. (1978) 'Differences in chemical composition and carbohydrate metabolism of *Echinococcus granulosus* (horse and sheep strains) and *E. multilocularis*' *Parasitology* 77, 103-109

McNeill, K.M. and Hutchinson, H.F. (1972) 'Carbohydrate metabolism of *Dirofilaria immitis*' in R.E. Bradley (ed.), *Canine Heartworm Disease: The Current Knowledge* (University of Florida Press, USA) pp 51-54

Mellin, T.N., Busch, R.D., Wang, C.C. and Kath, G. (1983) 'Neuropharmacology of the parasitic trematode *Schistosoma mansoni*' *American Journal of Tropical Medicine and Hygiene* 32, 83-93

Mendis, A.H.W. and Townson, S. (1985) 'Evidence for the occurrence of respiratory electron transport in adult *Brugia pahangi* and *Dipetalonema viteae*' *Molecular and Biochemical Parasitology* 14, 337-354

Mercer, J.G. (1985) 'Developmental hormones in parasitic helminths' *Parasitology Today* 1, 96-100

Meshnick, S.R. (1984a) 'The chemotherapy of African trypanosomes' in J. Mansfield (ed.) *Parasitic Diseases*. Vol. 2, 'The Chemotherapy' (M. Dekker, New York) pp 165-199

Meshnick, S.R. (1984b) 'Recent studies on inhibitors of macromolecular synthesis and function in trypanosomes' *Pharmacology and Therapeutics* 25, 239-254

Mettrick, D.F. and Podesta, R.B. (1974) 'Ecological and physiological aspects of helminth-host interactions in the mammalian gastrointestinal canal' *Advances in Parasitology* 12, 183-278

Mettrick, D.F. and Rahman, M.S. (1984) 'Effects of parasite strain and intermediate host species on carbohydrate intermediary metabolism in the rat tapeworm, *Hymenolepis diminuta*' *Canadian Journal of Zoology* 62, 355-361

Michels, P.A.M., Michels, J.P.J., Boonstra, J. and Konings, W.N. (1979) 'Generation of an electrochemical proton gradient in bacteria by the excretion of metabolic end-products' *FEMS Microbiology Letters* 5 357-364

Miles, M.A. (1983) '*Trypanosoma* and *Leishmania*: the contribution of enzyme studies to epidemiology and taxonomy' in G.S. Oxford and D. Rollinson (eds), *Protein Polymorphism: Adaptive and Taxonomic Significance* (Academic Press, London) pp 37-57

Miller, H.R.P. (1984) 'The protective mucosal response against gastrointestinal nematodes in ruminants and laboratory animals' *Veterinary Immunology and Immunopathology* 6, 167-259

Miller, H.R.P. (1986) 'Vaccination against intestinal parasites' in M.J. Howell (ed.), *Parasitology – Quo Vadit?* (Australian Academy of Science), pp 43-51

Mims, C.A. (1982) 'Innate immunity to parasitic infections' in S. Cohen and K.S. Warren (eds), *Immunology of Parasitic Infections* (Blackwell, Oxford), pp 3-27

Mitchell, G.F. (1979a) 'Responses to infection with metazoan and protozoan parasites in mice' *Advances in Immunology* 28, 451-511

Mitchell, G.F. (1979b) 'Effector cells, molecules and mechanisms in host-protective immunity to parasites' *Immunology* 38, 209-223

Mitchell, G.F. (1979c) 'Responses to infection with metazoan and protozoan parasites in mice' *Advances in Immunology* **38**, 209-223

Mitchell, G.F. (1982a) 'Effector mechanisms of host-protective immunity to parasites and evasion by parasites' in D.F. Mettrick and S.S. Desser (eds), *Parasites – Their World and Ours* (Elsevier, Amsterdam), pp 24-33

Mitchell, G.F. (1982b) 'New trends towards vaccination against parasites' *Clinics in Immunology and Allergy* **2**, 721-737

Mitchell, G.F. (1982c) 'Host-protective immune responses to parasites and evasion by parasites: generalizations and approaches to analysis' in L.E.A. Symons, A.D. Donald and J.K. Dineen (eds), *Biology and Control of Endoparasites* (Academic Press, Sydney), pp 331-341

Mitchell, G.F. and Anders, R.F. (1981) 'Parasite antigens and their immunogenicity in infected hosts' in M. Sela (ed.), *The Antigens* (Academic Press), p. 70

Mitchell, G.F. and Handman, E. (1977) 'Studies on the immune response to larval cestodes in mice: a simple mechanism of non-specific immunosuppression in *Mesocestoides corti* infected mice' *Australian Journal of Experimental Biology and Medical Science* **55**, 616-622

Mitchell, G.F. and Handman, E. (1985) 'T-lymphocytes recognise *Leishmania* glycoconjugates' *Parasitology Today* **1**, 61-63

Mitchell, G.F., Anders, R.F., Brown, G.V., Handman, L., Roberts-Thomson, I.C., Chapman, C.B., Forsyth, K.P., Kahl, L.P. and Cruise, K.M. (1982) 'Analysis of infection characteristics and antiparasite immune responses in resistant compared with susceptible hosts' *Immunological Reviews* **61**, 137-188

Moreno, S.N.J., Mason, R.P. and Docampo, R. (1984) 'Reduction of nifurtimox and nitrofurantoin to free radical metabolites by rat liver mitochondria. Evidence of an outer membrane-located nitroreductase' *Journal of Biological Chemistry* **259**, 6298-6305

Mukkada, A.J., Meade, J.C., Glaser, T.A. and Bonventre, P.F. (1985) 'Enhanced metabolism of *Leishmania donovani* amastigotes at acid pH: an adaptation for intracellular growth' *Science* **229**, 1099-1101

Müller, M. (1976) 'Carbohydrate and energy metabolism of *Tritrichomonas foetus*' in H. Van den Bossche (ed.), *Biochemistry of Parasites and Host-Parasite Relationships* (North Holland, Amsterdam) pp 3-4

Müller, M. (1980) 'The hydrogenosome' in G.W. Gooday, D. Lloyd and A.P.J. Trinci (eds), *The Eukaryotic Microbial Cell*. 30th Symposium od the Society of General Microbiology (Cambridge UP) pp 127-142

Muller, M. (1986) 'Reductive activation of nitroimidazole in anaerobic microorganisms' *Biochemical Pharmacology* **35**, 37-41

Müller, M. and Gorrell, T.E. (1983) 'Metabolism and metronidazole uptake in *Trichomonas vaginalis* isolates with different metronidazole susceptibilities' *Antimicrobial Agents and Chemotherapy* **24**, 667-673

Munir, W.A. and Barrett, J. (1985) 'The metabolism of xenobiotic compounds by *Hymenolepis diminuta* (Cestoda : Cyclophyllidea)' *Parasitology* **91**, 145-156

Myler, P.A., Nelson, R.G., Agabian, N. and Stuart, K. (1984) 'Two mechanisms of expression of a predominant variant antigen gene of *Trypanosoma brucei.*' *Nature, London* **309**, 282-284

Nelson, N.F. and Saz, H.J. (1982) 'Effects of levamisole on glycogen phosphorylase activity of *Litomosoides carinii*' *Journal of Parasitology* **68**, 1162-1163

Nevo, E. (1983) 'Adaptive significance of protein variation' in G.S. Oxford and D. Rollinson (eds) *Protein Polymorphism: Adaptive and Taxonomic Significance* (Academic Press, London) pp 239-282

Newbold, C.I. (1985) 'Parasite antigens in protection, diagnosis and escape. *Plasmodium*' *Current Topics in Microbiology and Immunology* 120, 69-104

Newsholme, E.A., Challis, R.A.J. and Crabtree, B. (1984) 'Substrate cycles – their role in improving sensitivity in metabolic control' *Trends in Biochemical Sciences* 9, 277-280

Newsholme, E.A. (1978) 'Control of energy provision and utilization in muscle in relation to sustained exercise' in F. Landry and W.A.R. Orban (eds) *3rd International Symposium on the Biochemistry of Exercise*, (Symposia Specialists, Miami) pp 3-27

Nolan, L.L., Berman, J.D. and Giri, L. (1984) 'The effect of formycin B on mRNA translation and uptake of purine precursors in *Leishmania mexicana*' *Biochemistry International* 9, 207-218

Nolan, T.J. and Farrell, J.P. (1985) 'Inhibition of *in vivo* and *in vitro* infectivity of *Leishmania donovani* by tunicamycin' *Molecular and Biochemical Parasitology* 16, 127-135

Nussenzweig, R.S. and Nussenzweig, V. (1985) 'Development of a sporozoite vaccine' *Parasitology Today* 1, 150-152

Nwagwu, M. and Opperdoes, F.R. (1982) 'Regulation of glycolysis in *Trypanosoma brucei*: hexokinase and phosphofructokinase activity' *Acta Tropica* 39, 61-72

Ogilvie, B.M. and De Savigny, D. (1982) 'Immune responses to nematodes' in S. Cohen and K.S. Warren (eds), *Immunology of Parasitic Infections* (Blackwell, Oxford), pp 715-757

Opperdoes, F.R. (1987) 'Compartmentation of carbohydrate metabolism in trypanosomes' *Annual Review of Microbiology* 41, 127-151

Opperdoes, F.R. and Cottem, D. (1982) 'Involvements of glycosome of *Trypanosoma brucei* in carbon dioxide fixation' *FEBS Letters* 143, 60-64

Opperdoes, F.R. and Van Roy, J. (1983) 'Involvement of lysosomes in the uptake of macromolecular material by bloodstream forms of *Trypanosoma brucei*' *Molecular and Biochemical Parasitology* 6, 181-190

Opperdoes, F.R., Misset, O. and Hart, D.T. (1984) 'Metabolic pathways associated with the glycosomes (microbodies) of the Trypanosomatidae' in J.T. August (ed.), *Molecular Parasitology* (Academic Press, New York) pp 63-75

Osoba, D., Dick, H.M., Voller, A., Goosen, T.J., Goosen, T., Draper, C.C. and Guy, De T. (1979) 'Role of HLA complex in the antibody response to malaria under natural conditions' *Immunogenetics* 8, 323-338

Ossikovski, E. and Walter, R.D. (1984) 'Cyclic AMP-dependent and independent protein kinases in *Ascaridia galli*' *Molecular and Biochemical Parasitology* 12, 299-306

Outteridge, P.M., Windon, R.G., Dineen, J.K. and Dawkins, H.J.S. (1984) 'Associations between lymphocyte antigens in sheep and responsiveness to vaccination against internal parasites' in J.K. Dineen and P.M. Outteridge (eds), *Immunogenetic Approaches to the Control of Endoparasites with Particular Reference to Parasites of Sheep* (CSIRO Publications), pp 103-111

Ovington, K.S. and Bryant, C. (1981) 'The role of carbon dioxide in the formation of end-products by *Hymenolepis diminuta*' *International Journal for Parasitology* 11, 221-228

Ovington, K.S., Barcarese-Hamilton, A.J. and Bloom, S.R. (1985) '*Nippostrongylus brasiliensis*: changes in plasma levels of gastrointestinal hormones in the infected rat' *Experimental Parasitology* 60, 276-284

Owen, D.G. (1982) (ed.) *Animal Models in Parasitology* (MacMillan, New York)

Oya, H. and Kita, H. (1989) The physiological significance of complex II (succinate-ubiquinone reductase) in respiratory adaptation' in E.M. Bennet, C.A. Behm and C. Bryant (eds) *Comparative Biochemistry of Parasitic Helminths* (Chapman and Hall, London) pp 35-55

Oya, H., Costello, L.C. and Smith, W.N. (1963) 'The comparative biochemistry of developing *Ascaris* eggs. II. Changes in cytochrome c oxidase activity during embryonation' *Journal of Cellular and Comparative Physiology* **62**, 287-294

Paltauf, F. and Meingassner, J.G. (1982) 'The absence of cardiolipin in hydrogenosomes of *Trichomonas vaginalis* and *Tritrichomonas foetus* ' *Journal of Parasitology* **68**, 949-950

Pappas, P.A. and Leiby, D.A. (1986a) 'Variation in the sizes of eggs and oncospheres and the numbers and distributions of testes in the tapeworm, *Hymenolepis diminuta*' *Journal of Parasitology* **72**, 383-391

Pappas, P.W. and Leiby, D.A. (1986b) 'Alkaline phosphatase and phosphodiesterase activities of the brush border membrane of four strains of the tapeworm, *Hymenolepis diminuta*' *Journal of Parasitology* **72**, 809-811

Pappas, P.W. and Read, C.P. (1975) 'Membrane transport in parasitic helminths: a review' *Experimental Parasitology* **37**, 469-530

Parkhouse, R.M.E. (1985) (ed.), 'Parasite antigens in protection, diagnosis and escape' *Current Topics in Microbiology and Immunology* **120**, 1-260

Parsons, M., Nelson, R.G. and Agabian, N. (1984) 'Antigenic variation in African trypanosomes: DNA rearrangements program immune evasion' *Immunology Today* **5**, 43-50

Pattanakitsakul, S-N. and Ruenwongsa, P. (1984) 'Characterization of thymidylate synthetase and dihydrofolate reductase from *Plasmodium berghei*' *International Journal for Parasitology* **14**, 513-520

Patthey, J-P. and Deshusses, J. (1987) 'Accessibility of *Trypanosoma brucei* procyclic glycosomal enzymes to labelling agents of various sizes and charges' *FEBS Letters* **210**, 137-141

Pearson, R.D., Manian, A.A., Harcus, J.L., Hall, D. and Hewlett, E.L. (1982) 'Lethal effect of phenothiazine neuroleptics on the pathogenic protozoan *Leishmania donovani*.' *Science* **217**, 369-371

Pegg, A.E. and McCann, P.P. (1982) 'Polyamine metabolism and function' *American Journal of Physiology* **243**, C212-C221

Penketh, P.G. and Klein, R.A. (1986) 'Hydrogen peroxide metabolism in *Trypanosoma brucei*' *Molecular and Biochemical Parasitology* **20**, 111-121

Pennak, R.W. (1978) *Freshwater Invertebrates of the United States*. 4th ed. (Wiley, New York)

Perlmann, P. (1986) 'Immunogenicity assays for clinical trials of malaria vaccines' *Parasitology Today* **2**, 127-130

Peters, W. (1985) 'The problem of drug resistance in malaria' *Parasitology* **90**, 705-715

Pfaller, M.A. and Krogstad, D.J. (1983) 'Oxygen enhances the antimalarial activity of the imidazoles' *American Journal of Tropical Medicine and Hygiene* **32**, 660-665

Piessens, W.F. and Mackenzie, C.D. (1982) 'Immunology of lymphatic filariasis and onchocerciasis' in D.F. Mettrick and S.S. Desser (eds), *Parasites – Their World and Ours* (Elsevier, Amsterdam) pp 622-653

Podesta, R.B. (1977) '*Hymenolepis diminuta*: unstirred layer thickness and effects on active and passive transport kinetics' *Experimental Parasitology* **43**, 12-34

Podesta, R.B. and Mettrick, D.F. (1974) 'Pathophysiology of cestode infections: effect of *Hymenolepis diminuta* on oxygen tensions, pH and gastrointestinal function' *International Journal for Parasitology* **4**, 278-292

Podesta, R.B. and Mettrick, D.F. (1976) 'The inter-relationships between the *in situ* fluxes of water, electrolytes and glucose by *Hymenolepis diminuta*' *International Journal for Parasitology* **6**, 163-172

Polak, A. and Richle, R. (1978) 'Mode of action of the 2-nitroimidazole derivative benznidazole' *Annals of Tropical Medicine and Parasitology* **72**, 45-54

Pong, S.S. and Wang, C.C. (1982) 'Avermectin B_1a modulation of gamma-aminobutyric acid receptors in rat brain membranes' *Journal of Neurochemistry* **38**, 375-379

Price, P.W. (1980) *Evolutionary Biology of Parasites* (Princeton University Press, Princeton)

Prichard, R.K., Bachmann, R., Hutchinson, G.W. and Köhler, P. (1982) 'The effect of praziquantel on calcium in *Hymenolepis diminuta*' *Molecular and Biochemical Parasitology* **5**, 297-308

Racagni, G.E. and Machedo de Domenech, E.E. (1983) 'Characterization of *Trypanosoma cruzi* hexokinase' *Molecular and Biochemical Parasitology* **9**, 181-188

Rainey, P. and Santi, D.V. (1984) 'Formycin B resistance in *Leishmania*' *Biochemical Pharmacology* **33**, 1374-1377

Ramp, T. and Köhler, P. (1984) 'Glucose and pyruvate catabolism in *Litomosoides carinii*' *Parasitology* **89**, 229-244

Ramp, T., Bachmann, R. and Köhler, P. (1985) 'Respiration and energy conservation in the filarial worm, *Litomosoides carinii*' *Molecular and Biochemical Parasitology* **15**, 11-20

Rangel-Aldao, R. and Opperdoes, F.R. (1984) 'Subcellular distribution and partial characterization of the cyclic AMP-binding proteins of *Trypanosoma brucei*' *Molecular and Biochemical Parasitology* **10**, 231-241

Rangel-Aldao, R., Allende, O. and Cayama, E. (1985) 'A unique type of cyclic AMP-binding protein of *Trypanosoma cruzi*' *Molecular and Biochemical Parasitology* **14**, 75-82

Rathod, P.K. and Reyes, P. (1983) 'Orotidylate-metabolizing enzymes of the human malarial parasite, *Plasmodium falciparum*, differ from host cell enzymes' *Journal of Biological Chemistry* **258**, 2852-2855

Raventos-Suarez, C., Pollack, S. and Nagel, R.L. (1982) '*Plasmodium falciparum* : inhibition of *in vitro* growth by desferrioxamine' *American Journal of Tropical Medicine and Hygiene* **31**, 919-922

Read, C.P. (1956) 'Carbohydrate metabolism of *Hymenolepis diminuta*' *Experimental Parasitology* **5**, 325-344

Reeves, R.E. (1984a) 'Metabolism of *Entamoeba histolytica* Schaudinn, 1903' *Advances in Parasitology* **23**, 105-142

Reeves, R.E. (1984b) 'Pyrophosphate energy conservation in a parasite and its suppression by chemical agents' in J.T. August (ed.), *Molecular Parasitology* (Academic Press, Orlando) pp 267-282

Rew, R.S. and Fetterer, R.H. (1984) 'Effects of diamfenetide on metabolic and excretory functions of *Fasciola hepatica in vitro*' *Comparative Biochemistry and Physiology* **79C**, 353-356

Rew, R.S., Smith, C. and Colglazier, M.L. (1982) 'Glucose metabolism of *Haemonchus contortus* adults: effects of thiabendazole on susceptible versus resistant strain' *Journal of Parasitology* **68**, 845-850

Richards, F.F. (1984) 'The surface of the African trypanosomes' *Journal of Protozoology* **31**, 60-64.

Rickard, M.D. and Howell, M.J. (1982) 'Comparative aspects of immunity in fascioliasis and cysticercosis in domesticated animals' in L.E.A. Symons, A.D. Donald and J.K. Dineen (eds), *Biology and Control of Endoparasites* (Academic Press, Sydney) pp 343-373

Rickard, M.D. and Williams, J.F. (1982) 'Hydatidosis/cysticercosis: immune mechanisms and immunisation against infection' *Advances in Parasitology* 21, 229-297

Rickman, L.R., Ernest, A., Dukes, P. and Maudlin, I. (1984) 'The acquisition of human serum resistance during cyclical passage of a *Trypanosoma brucei brucei* clone through *Glossina morsitans morsitans* maintained on human serum' *Transactions of the Royal Society of Tropical Medicine and Hygiene* 78, 284

Rifkin, M.R. and Fairlamb, A.H. (1985) 'Transport of ethanolamine and its incorporation into the variant surface glycoprotein of bloodstream forms of *Trypanosoma brucei*' *Molecular and Biochemical Parasitology* 15, 245-256

Roitt, I. (1984) *Essential Immunology* (Fifth ed.) (Blackwell, Oxford).

Roveri, O.A., de Cazzulo, B.M.F. and Cazzulo, J.J. (1982) 'Inhibition by suramin of oxidative phosphorylation in *Crithidia fasciculata*' *Comparative Biochemistry and Physiology* 71B, 611-616

Ruben, L. and Patton, C.L. (1985) 'Comparative structural analysis of calmodulins from *Trypanosoma brucei, T. congolense, T. vivax, Tetrahymena thermophila* and bovine brain' *Molecular and Biochemical Parasitology* 17, 331-341

Ruben, L., Egwuagu, C. and Patton, C.L. (1983) 'African trypanosomes contain calmodulin which is distinct from host calmodulin' *Biochimica et Biophysica Acta* 758, 104-113

Salzman, T.A., Stella, A.M., Wider de Xifra, E.A., Batlle, A.M., Docampo, R. and Stoppani, A.O.M. (1982) 'Porphyrin biosynthesis in parasitic hemoflagellates: functional and defective enzymes in *Trypanosoma cruzi*' *Comparative Biochemistry and Physiology* 72B, 663-668

Sani, B.P. and Comley, J.C.W. (1985). 'Rolke of retinoids and their binding proteins in filarial parasites and host tissues'.*Tropenmedizin und Parasitologie* 36 (supplement), 20-23

Saz, H.J. and Dunbar, G.A. (1975) 'The effects of stibophen on phosphofructokinases and aldolases of adult filariids' *Journal of Parasitology* 61, 794-801

Schad, G. (1966) 'Immunity, competition and natural regulation of helminth populations' *American Naturalist* 100, 359-364

Schirmer, R.H., Lederbogan, F., Eisenbrand, G. and Königk, E. (1984) 'Inhibitors of glutathione reductase as potential antimalarial drugs' *Parasitology* 89, i

Schnell, S., Becker, W. and Winkler, A. (1985) 'Amino acid metabolism in the freshwater pulmonate *Biomphalaria glabrata* infected with the trematode *Schistosoma mansoni*' *Comparative Biochemistry and Physiology* 81B, 1001-1008

Schopf, T.J.M. (1982) 'A critical assessment of punctuated equilibria. I. Duration of taxa' *Evolution* 36, 114-115

Schulman, M.D., Ostlind, D.A. and Valentino, D. (1982) 'Mechanism of action of MK-401 against *Fasciola hepatica*: inhibition of phosphoglycerate kinase' *Molecular and Biochemical Parasitology* 5, 133-145

Schwartz, A.L. and Hollingdale, M.R. (1985) 'Primaquine and lysosomotropic amines inhibit malaria sporozoite entry into human liver cells' *Molecular and Biochemical Parasitology* 14, 305-311

Scott, D.A., Coombs, G.H. and Sanderson, B.E. (1984) 'Folate metabolism in *Leishmania mexicana*' *Parasitology* 89, ii

Scott, M.T. and Snary, D. (1982) 'American trypanosomiasis (Chagas' disease)' in D.F. Mettrick and S.S. Desser (eds), *Parasites – Their World and Ours* (Elsevier, Amsterdam), pp 261-298

Scott, P. (1985) 'Impaired macrophage leishmanicidal activity at cutaneous temperature' *Parasite Immunology* 7, 277-288

Seebeck, T. and Gehr, P. (1983) 'Trypanocidal action of neuroleptic phenothiazines in *Trypanosoma brucei* *Molecular and Biochemical Parasitology* **9**, 197-208

Seed, J.L. and Bennett, J.L. (1980) '*Schistosoma mansoni*: phenol oxidase's role in eggshell formation' *Experimental Parasitology* **49**, 430-441

Seed, J.R., Edwards, R. and Sechelski, J. (1984) 'The ecology of antigenic variation' *Journal of Protozoology* **31**, 48-53

Selkirk, M.E., Denham, D.A., Partono, F., Sutanto, I. and Maizels, R.M.(1986) 'Molecular characterisation of antigens of lymphatic filarial parasites' *Parasitology* **91**, 515-538.

Senft, A.W. and Crabtree, G.W. (1983) 'Purine metabolism in the schistosomes: potential targets for chemotherapy' *Pharmacology and Therapeutics* **20**, 341-356

Sethi, K.K. (1982) 'Intracellular killing of parasites by macrophages' *Clinics in Immunology and Allergy* **2**, 541-565

Sharma, S., Svec, P., Mitchell, G.H. and Godson, G.N. (1985) 'Diversity of circumsporozoite antigen genes from two strains of the malarial parasite, *Plasmodium knowlesi*' *Science* **229**, 779-782

Sher, A. and Scott, P.A. (1982) 'Genetic factors influencing the interaction of parasites with the immune system' *Clinics in Immunology and Allergy* **2**, 489-510

Sherman, I.W. (1983) 'Metabolism and surface transport of parasitized erythrocytes in malaria' in E.J. Whelan (ed.), *Malaria and the Red Cell*, Ciba Foundation Symposium, Vol. 94 (Pitman, London) pp 206-216

Sherman, I.W. (1984) 'Metabolism' in W. Peters and W.H.G. Richards. (eds), *Antimalarial Drugs*. I. "Biological Background, Experimental Methods, and Drug Resistance." Handbook of Experimental Pharmacology, Vol. 68, No. 1 (Springer-Verlag, Berlin) pp 31-81

Siddiqui, A.A., Karcz, S.R. and Podesta, R.B. (1987) 'Developmental and immune regulation of gene expression in *Hymenolepis diminuta*' *Molecular and Biochemical Parasitology* **25**, 19-28

Simpson, A.J.G. and Cioli, D. (1987) 'Progress towards a defined vaccine for schistosomiasis' *Parasitology Today* **3**, 26-28

Sirawaraporn, W. and Yuthavong, Y. (1984) 'Kinetic and molecular properties of dihydrofolate reductase from pyrimethamine-sensitive and pyrimethamine-resistant *Plasmodium chabaudi*' *Molecular and Biochemical Parasitology* **10**, 355-367

Smith, C.K. and Strout, R.G. (1980) '*Eimeria tenella*: effect of narasin, a polyether antibiotic, on the ultrastructure of intracellular sporozoites' *Experimental Parasitology* **50**, 426-436

Smith, N.C. and Bryant, C. (1986) 'The role of host generated free radicals in helminth infections: *Nippostrongylus brasiliensis* and *Nematospiroides dubius* compared' *International Journal for Parasitology* **16**, 617-622

Smithers, S.R. (1976) 'Immunity to trematode infections' in S. Cohen and E. Sadun (eds), *Immunology of Parasitic Infections* (Blackwell, Oxford), pp 296-332.

Smithers, S.R. (1986) 'Vaccination against schistosomes and other systemic helminths' in M.J. Howell (ed.), *Parasitology – Quo Vadit?* (Australian Academy of Science), pp 30-40

Smithers, S.R. and Doenkoff, M.J. (1982) 'Schistosomiasis' in S. Cohen and K.S. Warren (eds), *Immunology of Parasitic Infections* (Blackwell, Oxford), pp 527-607

Smyth, J.D. (1962) 'Lysis of *Echinococcus granulosus* by surface active agents in bile and the role of this phenomenon in determining host specificity in helminths' *Proceedings of the Royal Society of London, Series B* **156**, 533-572

Smyth J.D. (1969) 'Parasites as biological models' *Parasitology* **59**, 73-91

Steinart, M. and Pays, E. (1986) 'Selective expression of surface antigen genes in African trypanosomes' *Parasitology Today* **2**, 15-19

Steinbüchel, A. and Müller, M. (1986a) 'Glycerol, a metabolic end product of *Trichomonas vaginalis* and *Tritrichomonas foetus*' *Molecular and Biochemical Parasitology* **20**, 45-55

Steinbüchel, A. and Müller, M. (1986b) 'Anaerobic pyruvate metabolism of *Tritrichomonas foetus* and *Trichomonas vaginalis* hydrogenosomes' *Molecular and Biochemical Parasitology* **20**, 57-65

Tait, A., Babiker, E.A. and Le Ray, D. (1984) 'Enzyme variation in *Trypanosoma brucei* ssp. I. Evidence for the sub-speciation of *Trypanosoma brucei gambiense*' *Parasitology* **89**, 311-326

Tait, A., Barry, J.D., Wink, R., Sanderson, A. and Crowe, J.S. (1985) 'Enzyme variation in *Trypanosoma brucei* ssp. II. Evidence for *T. b. rhodesiense* being a set of variants of *T. b. brucei* *Parasitology* **90**, 89-100

Taliafero, W.H. (1932) 'Trypanocidal and reproduction-inhibiting antibodies to *Trypanosoma lewisi* in rats and rabbits' *American Journal of Hygiene* **16**, 32-84

Targett, G.A.T. and Sinden, R.E. (1985) 'Transmission blocking vaccines' *Parasitology Today* **1**, 155-158

Taylor, A.E.R. and Baker, J.R. (1987) *In vitro Methods for Parasite Cultivation.* (Academic Press, London)

Taylor, M.B. and Gutteridge, W.E. (1986) 'The regulation of phosphofructokinase in epimastigote *Trypanosoma cruzi*' *FEBS Letters* **210**, 262-266

Téllez-Iñón, M., Ulloa, R.M., Torruella, M. and Torres, H.N. (1985) 'Calmodulin and Ca^{2+}-dependent cyclic AMP phosphodiesterase activity in *Trypanosoma cruzi*' *Molecular and Biochemical Parasitology* **17**, 143-153

Thompson, J.N. (1982) *Interaction and Coevolution.* (Wiley, New York)

Thompson, R.C.A. and Lymbery, A.J. (1988) 'The nature, extent and significance of variation within the genus *Echinococcus*' *Advances in Parasitology* **27**, 210-258

Thong, K-W. and Coombs, G.H. (1987) 'Comparative study of ferredoxin-linked and oxygen-metabolising enzymes of trichomonads' *Comparative Biochemistry and Physiology* **87B**, 637-641

Thorne, K.J.I. and Blackwell, J.M. (1983) 'Cell mediated killing of protozoa' *Advances in Parasitology* **22**, 43-151

Tielens, A.G.M., Van den Heuvel, J.M. and Van den Bergh, S.G. (1984) 'The energy metabolism of *Fasciola hepatica* during its development in the final host' *Molecular and Biochemical Parasitology* **13**, 301-307

Tielens, A.G.M. and Hill, G.C. (1985) 'The solubilization of a SHAM sensitive, cyanide insensitive ubiquinol oxidase from *Trypanosoma brucei*' *Journal of Parasitology* **71**, 384-386

Tielens, A.G.M., van der Meer, P. and Van Den Bergh, S.G. (1981) 'The aerobic energy metabolism of the juvenile *Fasciola hepatica*' *Molecular and Biochemical Parasitology* **3**, 205-214

Tracy, J.W., Catto, B.A. and Webster, L.T. (1983) 'Reductive metabolism of niridazole by adult *Schistosoma mansoni*. Correlation with covalent drug binding' *Pharmacology* **24**, 291-299

Turner, G. and Müller, M. (1983) 'Failure to detect extranuclear DNA in *Trichomonas vaginalis* and *Tritrichomonas foetus*' *Journal of Parasitology* **69**, 234-236

Turner, M.J. (1982) 'Biochemistry of the variant surface glycoproteins of salivarian trypanosomes' *Advances in Parasitology* **21**, 69-153

Turner, M.J. (1984) 'Antigenic variation in parasites' *Parasitology* **88**, 613-621

Turrens, J.F., Bickar, D. and Lehninger, A.L. (1986) 'Inhibitors of the mitochondrial b-c_1 complex inhibit the cyanide-insensitive respiration of *Trypanosoma brucei*' *Molecular and Biochemical Parasitology* 19, 259-264

Tuttle, J.V. and Krenitsky, T.A. (1980) 'Purine phosphoribosyl transferases from *Leishmania donovani*' *Journal of Biological Chemistry* 255, 909-916

Uglem, G.L., Lewis, M.C. and Larson, O.R. (1985) 'Niche segregation and sugar transport capacity of the tegument in digenean flukes' *Parasitology* 91, 121-128

Van den Bossche, H. (1985) 'How anthelmintics help us to understand helminths' *Parasitology* 90, 675-686

Van den Bossche, H. and Verhoeven, H. (1982) 'Biochemical effects of the antiparasitic drug closantel' *Parasitology* 84, li

Van der Jagt, D.L., Hunsaker, L.A. and Heidrich, J.E. (1981) 'Partial purification and characterization of lactate dehydrogenase from *Plasmodium falciparum*' *Molecular and Biochemical Parasitology* 4, 255-264

Van Oordt, B.E.F., Van Den Heuvel, J.M., Tielens, A.G.M. and Van Den Bergh, S.G. (1985) 'The energy production of the adult *Schistosoma mansoni* is for a large part aerobic' *Molecular and Biochemical Parasitology* 16, 117-126

Vargas, F. Del C., Viens, P. and Kongshavn, P.A.L. (1984) '*Trypanosoma musculi* infection in B-cell-deficient mice' *Infection and Immunity* 44, 162-167

Verdier, F., LeBras, J., Clavier, F., Hatin, I. and Blayo, M.C. (1985) 'Chloroquine uptake by *Plasmodium falciparum*-infected human erythrocytes during *in vitro* culture and its relationship to chloroquine resistance' *Antimicrobial Agents and Chemotherapy* 27, 561-564

Vial, H.J., Thuet, M.J., Ancelin, M.L., Philippot, J.R. and Chavis, C. (1984) 'Phospholipid metabolism as a new target for malaria chemotherapy. Mechanism of action of D-2-amino-1-butanol' *Biochemical Pharmacology* 33, 2761-2770

Vial, H.J., Thuet, M.J., Broussal, J.L. and Philippot, J.R. (1982) 'Phospholipid biosynthesis by *Plasmodium knowlesi*-infected erythrocytes: the incorporation of phospholipid precursors and the identification of previously undetected metabolic pathways' *Journal of Parasitology* 68, 379-391

Vial, H.J., Torpier, G., Ancelin, M.L. and Capron, A. (1985) 'Renewal of the membrane complex of *Schistosoma mansoni* is closely associated with lipid metabolism' *Molecular and Biochemical Parasitology* 17, 203-218

Vickerman, K. (1978) 'Antigenic variations in trypanosomes' *Nature, London* 273, 613-617

Vickerman, K. and Barry, J.D. (1982) 'African trypanosomiasis' in S. Cohen and K.S. Warren (eds), *Immunology of Parasitic Infections* (Blackwell, Oxford), pp 204-260

Viens, P., Targett, G.A.T. and Lumsden, W.H.R. (1975) 'The immunological response of CBA mice to *Trypanosoma musculi*: mechanisms of protective immunity' *International Journal for Parasitology* 5, 235-239

Villalta, F. and Kierszenbaum, F. (1985) 'The effect of swainsonine on the association of *Trypanosoma cruzi* with host cells' *Molecular and Biochemical Parasitology* 16, 1-10

Wakelin, D. (1978a) 'Genetic control of susceptibility and resistance to parasitic infection' *Advances in Parasitology* 16, 219-308

Wakelin, D. (1978b) 'Genetic control of the immune response: survival within low responder individuals of the host population' *Parasitology* 88, 639-657

Wakelin, D. (1984) *Immunity to Parasites* (Edward Arnold, London)

Wakelin, D. (1986) 'The role of the immune response in helminth population regulation' in M.J. Howell (ed.), *Parasitology – Quo Vadit?* (Australian Academy of Science) pp 549-557

Waller, P.J. and Lacey, E (1985) 'Nematode growth regulators' in N. Anderson and P.J.Waller (eds.) *Resistance in Nematodes to Anthelmintic Drugs*. CSIRO Division of Animal Health/Australian Wool Corporationn Sydney, Australia, pp137-147

Walliker, D. (1983a) 'Genetics of parasites' in J. Guardiola, L. Luzzatto and W. Trager (eds) *Molecular Biology of Parasites* (Raven Press, New York) pp 53-62

Walliker, D. (1983b) 'Enzyme variation in malaria parasite populations' in G.S. Oxford and D. Rollinson (eds) *Protein Polymorphism: Adaptive and Taxonomic Significance* (Academic Press, London) pp 27-35

Walter, R.D. (1980) 'Effect of suramin on phosphorylation-dephosphorylation reactions in *Trypanosoma gambiense*' *Molecular and Biochemical Parasitology* 1, 139-142

Walter, R.D. and Albiez, E.J. (1984) 'Interaction of amoscanate with the cyclic AMP-phosphodiesterases from *Schistosoma mansoni, Onchocerca volvulus* and bovine brain' *Tropenmedizin und Parasitologie* 35, 78-80

Walter, R.D. and Albiez, E. (1985) 'Parasite-specific inhibition of 5'-nucleotidase from *Onchocerca volvulus* and *Dirofilaria immitis* by the amoscanate-derivative CGP 8065' *Molecular and Biochemical Parasitology* 16, 109-115

Walter, R.D. and Ossikovski, E. (1985) 'Effect of amoscanate derivative CGP 8065 on aminoacyl-tRNA synthetases in *Ascaris suum*' *Molecular and Biochemical Parasitology* 14, 23-28

Walter, R.D. and Schulz-Key, H. (1980) 'Interaction of suramin with protein kinase I from *Onchocerca volvulus*' in H. Van den Bossche (ed.), *The Host-Invader Interplay* (Elsevier, Amsterdam) pp 709-712

Walter, R.D., Ossikovski, E. and Albiez, E.J. (1985) 'Properties of dolichol kinase from *Onchocerca volvulus* and *Ascaris suum*' *Tropical Medicine and Parasitology* 36 (Suppl.), 29

Wang, A.L. and Wang, C.C. (1985) 'Isolation and characterization of DNA from *Tritrichomonas foetus* and *Trichomonas vaginalis*' *Molecular and Biochemical Parasitology* 14, 323-335

Wang, C.C. (1984) 'Parasite enzymes as potential targets for antiparasitic chemotherapy' *Journal of Medicinal Chemistry* 27, 1-9

Wang, C.C. and Cheng, H.W. (1984a) 'Salvage of pyrimidine nucleosides by *Trichomonas vaginalis*' *Molecular and Biochemical Parasitology* 10, 171-184

Wang, C.C. and Cheng, H.W. (1984b) 'The deoxyribonucleoside phosphotransferase of *Trichomonas vaginalis*. A potential target for anti-trichomonal chemotherapy' *Journal of Experimental Medicine* 160, 987-1000

Wang, C.C. and Simashkevich, P.M. (1981) 'Purine metabolism in the protozoan parasite *Eimeria tenella*' *Proceedings of the National Academy of Sciences*, USA 78, 6618-6622

Wang, C.C., Verham, R., Cheng, H.W., Rice, A. and Wang, A.L. (1984) 'Differential effects of inhibitors of purine metabolism on two trichomonad species' *Biochemical Pharmacology* 33, 1323-1329

Wang, C.C., Verham, R., Tzeng, S.F., Aldritt, S. and Cheng, H.W. (1983) 'Pyrimidine metabolism in *Tritrichomonas foetus*' *Proceedings of the National Academy of Sciences, USA* 80, 2564-2568

Warren, K.S. (1978) 'The pathology, pathobiology and pathogenesis of schistosomiasis' *Nature, London* 273, 609-612

Washtien, W.L., Grumont, R. and Santi, D.V. (1985) 'DNA amplification in antifolate-resistant *Leishmania*. The thymidylate synthase-dihydrofolate reductase' *Biological Chemistry* **260**, 7809-7812

Watkins, W.M., Sixsmith, D.G., Chulay, J.D. and Spenrer, H.C. (1985) 'Antagonism of sulfadoxine and pyrimethamine antimalarial activity *in vitro* by *p*-aminobenzoic acid, *p*-aminobenzoylglutamic acid and folic acid' *Molecular and Biochemical Parasitology* **14**, 55-61

Watts, S.D.M. and Fairbairn, D. (1974) 'Anaerobic excretion of fermentation acids by *Hymenolepis diminuta* during development in the definitive host' *Journal of Parasitology* **60**, 621-625

Weinbach, E.C. (1981) 'Biochemistry of enteric parasitic protozoa' *Trends in Biochemical Sciences* **6**, 254-257

Weinbach, E.C., Clagett, C.E., Keister, D.B., Diamond L.S. and Kon H. (1980) 'Respiratory metabolism of *Giardia lamblia*' *Journal of Parasitology* **66**, 347-330

Wharton, D.A. (1983) 'The production and functional morphology of helminth egg shells' *Parasitology* **86**(4), 85-97

Whaun, J.M., Brown, N.D. and Chiang, P.K. (1984) 'Effects of two methylthioadenosine analogues, SIBA and DEAZA-SIBA, on *P. falciparum*-infected red cells' in J.W. Eaton and G.J. Brewer (eds), *Malaria and the Red Cell*, Progress in Clinical and Biological Research Vol. 155, (Alan R. Liss, New York) pp 143-157

White, E., Hart, D. and Sanderson, B.E. (1983) 'Polyamines in *Trichomonas vaginalis*' *Molecular and Biochemical Parasitology* **9**, 309-318

Williams, J.F. (1982) 'Cestode infections' in S. Cohen and K.S. Warren (eds), *Immunology of Parasitic Infections* (Blackwell, Oxford), pp 676-714

Williams, K., Lowe, P.N. and Leadlay, P.F. (1987) 'Purification and characterization of pyruvate:ferredoxin oxidoreductase from the anaerobic protozoan *Trichomonas vaginalis*' *Biochemical Journal* **246**, 529-536

Wiser, M.F., Eaton, J.W. and Sheppard, J.R. (1983) 'A *Plasmodium* protein kinase that is developmentally regulated, stimulated by spermine and inhibited by quercetin' *Journal of Cellular Biochemistry* **21**, 305-314

Woo, P.T.K. (1981) 'Acquired immunity against *Trypanosoma danilewskyi* in goldfish, *Carassius auratus*' *Parasitology* **83**, 343-346

Wood, P.A., Rock, L.M. and Eaton, J.W. (1984) 'Chloroquine resistance and host cell hemoglobin catabolism in *Plasmodium berghei*' in J.W. Eaton and G.J. Brewer (eds), *Malaria and the Red Cell*, Progress in Clinical and Biological Research, Vol. 155, (Alan R.Liss, New York) pp 159-169

Wuntch, T. and Goldberg, E. (1969) 'A comparative physico-chemical characterisation of lactate dehydrogenase: isozymes in brook trout, lake trout and their hybrid lake trout' *Journal of Experimental Zoology* **174**, 233-252

Yarlett, N., Gorrell, T.E., Marczak, R. and Müller, M. (1985) 'Reduction of nitroimidazole derivatives by hydrogenosomal extracts of *Trichomonas vaginalis*' *Molecular and Biochemical Parasitology* **14**, 29-40

Yayon, A., Timberg, R., Friedman, S. and Ginsburg, H. (1984) 'Effects of chloroquine on the feeding mechanism of the intraerythrocytic human malarial parasite *Plasmodium falciparum*' *Journal of Protozoology* **31**, 367-372

Zilberstein, D. and Dwyer, D.M. (1985) 'Protonmotive force-driven active transport of D-glucose and L-proline in the protozoan parasite *Leishmania donovani*' *Proceedings of the National Academy of Science, USA* **82**, 1716-1720

Zingales, B., Katzin, A.M., Arruda, M.V. and Colli, W. (1985) 'Correlation of tunicamycin-sensitive surface glycoproteins from *Trypanosoma cruzi* with parasite interiorization into mammalian cells' *Molecular and Biochemical Parasitology* 16, 21-34

Zuckerman, A. (1963) 'Immunity in malaria with particular reference to red cell destruction' in P.C.C. Garnham, A.E. Pierce and I. Roitt (eds), *Immunity to Protozoa* (Blackwell, Oxford), pp 78-88

Zuckerman, A. (1970) 'Malaria of lower mammals' in G.J. Jackson, R. Herman and I. Singer (eds), *Immunity to Parasitic Animals*, Vol. 2, (Appleton-Century-Crofts, New York), pp 793-829

Zuk, M. (1984) 'A charming resistance to parasites' *Natural History* 4, 28-34

Index

Species index